Cognitive Visual Informatics

Hyunhee Kim

Education and Career

1985~present. Professor at the Myongji University (South Korea)

2003. 8~2004. 7. Visiting scholar at the University of Michigan (USA)

1985. Ph.D. in information science, Case Western Reserve University (USA)

1979. Master's degree in library science, Sungkyunkwan University (South Korea)

Publications

Kim, H., & Kim, Y. (2016). Generic speech summarization of transcribed lecture videos: Using tags and their semantic relations. Journal of the Association for Information Science and Technology, 67(2), 366~379.

Kim, H. (2011). Toward video semantic search based on a structured folksonomy. Journal of the Association for Information Science and Technology, 62(3), 478~492.

Kim, H., & Kim, Y. (2010). Toward a conceptual framework of key-frame extraction and storyboard display for video summarization. Journal of the Association for Information Science and Technology, 61(5), 927~939.

Cognitive Visual Informatics

Published in February 2018
Published by Kyung-in Publishing Company
Written by Hyunhee Kim

Kyung-in Publishing Company
Paju, Gyeonggi-do, 10881, Korea
Homepage: http://www.kyunginp.co.kr/

This work was supported by the Korea Research Foundation Grant funded by the Korean Government (NRF-2011-342-B00025)

ISBN 978-89-499-4726-6   93400
Price: 22,000 won
Printed in Korea

Cognitive Visual Informatics

Hyunhee Kim

Kyung-in Publishing Co.

Cognitive visual informatics is an interdisciplinary study of how humans organize, seek, store, and retrieve visual information. The practice of cognitive visual informatics research requires sophisticated theories, information technologies, and neurophysiological methods to design a natural user interface, manage information, and provide customized information services to users.

The book introduces theories and technologies that are applied to model and process visual information for this emerging field at the intersection of information science, library science, computer science, and cognitive neuroscience, which combines the research techniques of cognitive psychology with neuroscience techniques, such as electroencephalography/event−related potential (EEG/ERP), to assess the structure and function of the brain. With the rise of cognitive neuroscience, many people with no previous experience in electro−physiology began setting up their own ERP labs. This was an important trend, because these researchers brought considerable expertise from other areas of science and began applying ERPs to a broader range of issues. I was fortunate to be involved in EEG/ERP projects, which were supported by the National Research Foundation of Korea. My goal in writing this book was to summarize research results on cognitive visual informatics that were obtained from the projects.

The book seems to be different from other information science books in terms of applying neuroscience, and cognitive science to information science field. The volume is organized so as first to explain basic concepts and then to illustrate them with examples and case studies. Hence, the book is useful for researchers, students, and practitioners of information science and its areas of application who want to know the new trends and applications in this field.

2018. 2

Hyunhee Kim

C·O·N·T·E·N·T·S

Chapter 1

Introduction

1. Introduction

A large amount of visual information has been produced and accumulated. Nicola Mendelsohn, who heads up Facebook's operations in Europe, predicted that Facebook text will be replaced with photos and videos within the next five years, because visual information conveys far more information more quickly. Users who are familiar with text literacy do not easily grasp the meaning that multimedia information has. Furthermore, owing to a lack of proper surrogates and metadata, multimedia such as videos or speeches are often not used to their full potential as educational and academic support materials.

Wang (2003) defines cognitive informatics as a transdisciplinary study of cognitive and information sciences that investigates the internal information processing of the brain and the resulting engineering applications in the computing and information technology industry. Media informatics is defined as the study of how humans seek, use, manipulate, store, retrieve, and organize digital multimedia (Goodrum, Devereaux, Langlois, & Marchionini, 2009). Media informatics investigates the behaviors and practices related to new media objects and environments including the social, communication, and information aspects of new media content, the results of which can be applied to the design and development of tools for media access, retrieval, and storage.

The term "visual informatics" is used in various fields, such as computer science, cognitive psychology, and information science without a clear definition (Zaman et al., 2009). In other words, visual informatics encompasses various research areas including computer vision, information visualization, medical image processing, image search, virtual reality, visual ontology, visual data mining, visual culture, and visual services.

Cognitive visual informatics, which includes the "visual" aspect of cognitive informatics in order to focus on visual information processing, is defined as an interdisciplinary study of how humans organize, seek, store, and retrieve visual information, while dealing with sophisticated theories, information technologies, and neurophysiological technologies for modeling, representing, summarizing, and accessing visual information. Two points are emphasized in the definition of cognitive visual informatics. The first deals with a field of study focused on the users' point of view. It deals with how users acquire, store, transform, and use visual information, depending on different situations and how they interact with visual information retrieval systems. The second is the field of study focused on the technical point of view. In other words, it deals with visual information representation, summarization, and access by analyzing texts, colors, objects, and voice information.

Cognitive visual informatics is closely related to cognitive psychology and cognitive neuroscience as well as the abovementioned cognitive informatics and media informatics in terms of theories and methods. Cognitive psychology is the study of mental processes, such as attention, language use, memory, perception, problem solving, creativity and thinking (American Psychological Association); basically it examines on how humans think, perceive, learn, and remember. Cognitive neuroscience combines the research techniques of cognitive psychology with neuroscience techniques, such as electroencephalography (EEG), electrodermal activity (EDA), and functional magnetic resonance imaging (fMRI), to assess the structure and function of the brain.

Let us consider an example to examine how cognitive visual informatics, cognitive psychology, and cognitive neuroscience are related to one another. Suppose that there is a project whose goal is to determine how humans acquire, store, and transform visual information. In order to conduct the project, a user is asked to watch a video and to write down (or type) its topic after watching the video. The user needs to strategically allocate his/her attention to actions and events occurring in the video that seem topic-relevant in order to process auditory, linguistic, and visual information acquired from sensory systems.

Subsequently, the user links the acquired information to the knowledge stored in his/her working memory (WM) or long–term memory, creating an internal interpretation of the video. In the meantime, new memories of the user's experience can be created and then accessed at a later point in time when he/she writes down the topic of the video. In order to investigate the cognitive and emotional reactions of humans engaged in the abovementioned cognitive task process, we can adapt theories of cognitive psychology, cognitive neuroscience techniques (e.g., EEG), and self–reporting measure– ments (e.g., interview).

The research findings obtained from cognitive visual informatics can be applied to the design and operation of a digital library system or a natural user interface (NUI), leading to an improvement in the information representation and access efficiency. Specifically, proposed models and algorithms can be utilized to better understand the concept of relevance, and to extract topic–relevant semantic parts rather than perceptual visual information for the construction of a video skim that consists of the key shots extracted from the video. Additionally, these findings can be applied as basic data to develop users' visual literacy skills through which they can improve not only their access to visual information but also their ability to analyze and evaluate it for its relevance and usefulness to their tasks.

This book consists of nine chapters. Chapter 1 describes the concept of cognitive visual informatics and its related subjects. It also introduces an overview of the book. Chapter 2 explains the basic theories of cognitive visual informatics, such as relevance, as well as the theoretical framework for visual information recognition and processing, and Chatman's narrative theory. Chapter 3 introduces neuro−physiological (NP) studies on cognitive informatics, to describe and explain the generation of NP evidence using three types of modalities: eye tracking, EEG, and fMRI. The case study titled, "Understanding topical relevance judgment in visual simple search: An EEG study" written by the author is also described.

Chapter 4 introduces Baddeley's WM theory and Mayer's cognitive model of multimedia learning. Then, it explains the EEG/event related potential (ERP)−based relevance model that was proposed by the author using the abovementioned theories. Chapter 5 introduces three multimedia metadata formats (PBCore, MPEG−7, and TV−Anytime (TVA)), and also details the metadata framework for efficient browsing and searching of Web videos. This chapter is based on the paper "Toward a structural and semantic metadata framework for efficient browsing and searching of Web videos" which appeared in the Journal of the Korean Society for Library and Information Science, 51 (2016), 227~243, and is used by permission of the Korean Society for Library and Information Science.

Chapter 6 provides an overview of previous studies on three speech summarization methods, namely supervised, unsupervised, and social summarization methods. It also details four speech summariz‒ ation methods: social summarization, latent semantic analysis, maximum marginal relevance, and acoustic methods. Chapter 7 provides an overview of previous studies on video summarization methods, and also details two case studies, "Algorithm for key‒frame extraction" and "Video summarization using EEG/ERP techniques."

Chapter 8 describes social information retrieval in terms of its definition and its research areas, and introduces case studies on social indexing and search of videos. Chapter 9 describes multimedia retrieval focusing on the approach and performance evaluation. Then, it describes the evaluation result of the social summarization method that is introduced in Chapter 6, and that of the video summarization method using the EEG/ERP techniques that are described in Chapter 7.

References

American Psychological Association. (2013). Glossary of psychological terms, http://www.apa.org/research/action/glossary.aspx.

Bawden, D., & Robinson, L. (2015). Introduction to information science. Facet Publishing.

Bouadjenek, M. R., Hacid, H., & Bouzeghoub, M. (2016). Social networks and information retrieval, how are they converging? A survey, a taxonomy and an analysis of social information retrieval approaches and platforms. Information Systems, 56, 1~18.

Goodrum, A., Devereaux, Z., Langlois, G., & Marchionini, G. (2009). Media Informatics: Theory, methods, and tools. Proceedings of the American Society for Information Science and Technology, 46(1), 1~3.

Kim, H., & Kim, Y. (2016). Generic speech summarization of transcribed lecture videos: Using tags and their semantic relations. Journal of the Association for Information Science and Technology, 67(2), 366~379.

Mayer, R. E. (2005). Cognitive theory of multimedia learning. In R.E. Mayer (Ed.), The Cambridge handbook of multimedia learning (pp. 134~146). New York: Cambridge University Press.

Mostafa, J., & Gwizdka, J. (2016). Deepening the role of the user: Neuro-physiological evidence as a basis for studying and improving search. Proceedings of the 2016 ACM on Conference on Human Information Interaction and Retrieval (pp. 63~70). New York: ACM.

Saracevic, T. (2015). Why is relevance still the basic notion in information science. In Re: inventing Information Science in the Networked Society. Proceedings of the 14th International Symposium on Information Science (ISI 2015) (pp. 26~35).

Wang, Y. (2003). On cognitive informatics. Brain and Mind, 4(2), 151~167.

Zaman, H., Robinson, P., Petrou, M., Olivier, P., & Schröder, H. (Eds.). (2009). Visual informatics: Bridging research and practice: First

International Visual informatics Conference, IVIC 2009 Kuala Lumpur, Malaysia, November 11~13, 2009 Proceedings (Vol. 5857). Springer Science & Business Media.

Chapter 2

Theories

2. Theories

In this chapter, we explain the basic theories for cognitive visual informatics, namely relevance, the theoretical framework for visual information recognition and processing, and Chatman's narrative theory.

2.1 Relevance

The concept of relevance is studied in many different fields, including cognitive sciences and information retrieval. Different areas have different implications for what is considered relevant. In the context of information retrieval, there are two types of definitions for relevance: system−oriented relevance or user−oriented relevance (Yang & Marchionini, 2004). The system−oriented relevance concentrates on the relations between a specified search request and the retrieved documents, while the user−oriented relevance focuses on the relations between users' information needs and the retrieved documents. Many researchers have expressed this dichotomous view of relevance concept: for example, Swanson's (1986) objective relevance and subjective relevance. Objective relevance is crucial to the design and testing of information retrieval systems, whereas subjective relevance is paramount in the operation and use of such systems. Wilson (1973) mentioned that relevance is not a single notion, and the multidimensional nature of relevance has been agreed among researchers. For example, there are various types of relevance: situational relevance, psychological/cognitive relevance, dynamic relevance, and topical relevance. Relevance research is extensive (Gwizdka, Hosseini, Cole, & Wang, 2017), ranging from theoretical studies (Huang & Siegel, 2013; Saracevic, 2007), to behavioral studies (Fitzgerald & Galloway, 2001; Taylor, 2012), and to applied and system-oriented evaluation studies (Ruthven, 2014).

Topicality is always regarded as the most common and important criterion for relevance judgment, when searching for images and videos as well as textual information (Yang & Marchionini, 2004; Choi & Rasmussen, 2002). Topical relevance is defined as a process that determines whether the information at hand can serve as evidence for deriving reliable conclusions (Huang & Soergel, 2006). The topical relevance approach frames relevance judgment as a cognitive process, in which a retrieved item serves as a stimulus, leading to changes in a user's knowledge and information needs (Harter, 1992; Ingwersen, 1996; Saracevic, 2007; Chen & Xu,

2005)

Recent research has begun to utilize neuro−physiological (NP) methods to understand how topical relevance judgments during a search occur in the brain (Mostafa & Gwizdka, 2016). Allegretti et al. (2015) conducted an event−related potential (ERP)−based study on the first 800 ms of a relevance assessment process to determine the time at which topical relevance is assessed in the brain. Moshfeghi, Pinto, Pollick, and Jose (2013) conducted a functional magnetic resonance imaging study in order to understand the concept of topical relevance.

Although research on topical relevance has made progress, the process underlying topical relevance judgment is not well understood. This is because topical relevance judgment is a cognitive dynamic process influenced by many factors, including search task complexity (Huang & Soergel, 2013). For example, search tasks can be divided into two types: simple and complex (Komaki, Hara, & Nishio, 2012; Singer, Danilov, & Norbisrath, 2013). A simple search task (e.g., look−up searches) permits a search process in which a search topic is already clearly defined at the beginning of the search. Thus, the solution or answer to the simple search can be obtained with a query and result pairing.

On the contrary, the complex search task (e.g., exploratory searches) permits a search process in which a search topic is not clearly defined before the search; thus, the user needs to learn about the search topic in order to understand how to achieve his or her goal (Hendahewa & Shah, 2015). During the complex search process, the topic becomes clear as the information seeker aggregates, discovers, and synthesizes information (Singer, Danilov, & Norbisrath, 2012). As such, the complex search requires the user to undergo complex cognitive tasks that lead to exploring, discovering, and learning intellectual skills (Hendahewa & Shah, 2015; Hendahewa & Shah, 2017; Rieh, Collins−Thompson, Hansen, & Lee, 2016). Hence, we need to consider search task complexity while doing studies on topical relevance, and we will detail this issue in Chapter 4.3.

2.2 Theoretical Framework for Visual Information Recognition and Processing

According to Paivio's dual−coding theory (1986), visual and verbal information are processed differently and along distinct channels within the human mind, creating separate representations of the processed information. This means that information obtained via each channel plays its own unique role; the two reinforce each other, instead of competing with one another.

Several previous studies on video surrogates support Paivio's theory. Let us take a look at previous studies of the utility of a presentation style of video summaries that employ qualitative investigation of user cognitive processes. Ding, Marchionini, and Soergel (1999) examined three types of video summaries (visual, verbal [keywords/phrases], and joint visual and verbal). Their results showed that users favored the combined summary, in which verbal information and images reinforce each other; such information helped them to grasp the overall meaning of the video and specify or clarify the thematic material of the visual surrogate; in comparison, visual information was more apt to convey affect, emotion, and excitement and to draw attention. Turner (1994) suggestsed that text and images are complementary and interdependent aspects of video information.

Similarly, Hughes, Wilkins, Wildemuth, and Marchionini (2003) mentioned that verbal information can provide individuals with an understanding of things like video content, whereas visual information can provide individuals with a vivid and concrete understanding. Additionally, Wildemuth et al. (2002) argued that textual video surrogates can facilitate the process of determining relevance, and non-textual video surrogates can effectively augment textual surrogates.

Hughes et al. (2003) suggested that textual surrogates appear to transmit information regarding the contents of the video, while non-textual surrogates appear to transmit information regarding what the video was like. Yang and Marchionini (2004) found that users liked to see visual surrogates for relevance judgments, especially those surrogates that contained motion.

2.2.1 Uni-modal

Panofsky (1955) classified visual images into three categories: preiconography, iconography, and iconology. Shatford (1986) interpreted Panofsky's preiconography and iconography as generics and specifics, respectively. While the generics indicate the general subject matter of an image (e.g., bridge), the specifics indicate the specific subject matter of an image (e.g., Golden Gate Bridge). Furthermore, Shatford relates Panofsky's iconology to abstracts, which relate to the intrinsic, personal meaning of an image.

Eakins and Graham (1999) classified images into three levels of abstraction: primitive features (e.g., color), logical features (e.g., the identity of objects), and abstract features (e.g., the meaning of an image). Greisdorf and O'Connor (2002) suggested that people interact with images/videos at three levels. At the first level, the primitive features of an image are perceived, whereas at the second level, objects are identified. At the third level, inductive inter—pretation of an image/video is required, with inferences being made about its abstract attributes. As shown in Table 2–1, we classified each of the four studies into three categories: description, analysis, and interpretation.

According to the abovementioned studies, people interact with images at three levels. At the first level, they observe the primitive features of an image, such as shape. Then, at the second level, they

Table 2-1. Summary of previous studies on image recognition

Study \ Category	Description	Analysis	Interpretation
Panofsky (1955)	pre−iconography (factual, expressional)	iconography	iconology
Shatford (1986)	generics (object, event/ activity, time, space)	specifics (object, event/activity, time, space)	abstracts (object, emotion/abstraction, time, space)
Eakins & Graham (1999)	primitive features (color or textures, etc.)	logical features (the identity of objects)	abstract features (e.g., meaning or significance of an image)
Greisdorf & O'Connor (2002)	primitive features (color, shape, texture)	objects (person/ thing, place/ location, action (activity, event))	inductive interpretation (symbolic value, prototypical displacement (atmosphere, emotional cue))

consider derived attributes such as the presence of specific objects. Lastly, at the third level, they consider the semantic abstract attributes of the image, such as the symbolic value (Greisdorf & O'Connor, 2002).

Kim (2008) and Kim and Kim (2010) constructed a framework to investigate users' cognitive understanding about key–images (key–frames) in a video storyboard (see Figure 2–1). The framework was constructed based on the studies of Shatford (1986) and Yang and Marchionini (2004). It has the three categories (textual, visual, and implicit) identified by Yang and Marchionini (2004) but the textual category was redefined by giving it a narrower scope, applying

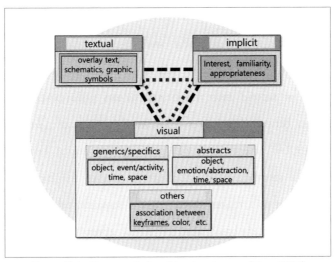

Figure 2-1. Framework for video key-frames (Kim, 2008)

it only to text or symbols in an image, such as overlay text and symbols. The visual category includes the generics, specifics, and abstracts suggested by Shatford (1986), and that also has associations between key frames, color, and others. The implicit category includes interest or familiarity.

Using this framework, Kim and Kim (2010) conducted a test in which 44 participants were asked to browse each of 12 video storyboard and to describe which frames in each storyboard were useful in identifying the content of the associated video and why. They obtained 132 responses; in 111 of the responses, only a single frame was identified as a key frame, while two or more frames were selected in the remaining 21 responses, primarily because of their having shared associations or relations.

Furthermore, "object" (26.5%) and "event/action" (24.2%) were found to be the most important factors in identifying the content of a video. The second most important factors were observed to be "association between key frames" (15.9%), "overlay text" (14.4%), and "person" (9.8%). Here, a viewer making associations between key frames on a storyboard, in an effort to extract the meaning of a video, is similar to a reader identifying the semantic relations between words within a text-based abstract, in an effort to understand the content of a document. None of the participants made reference to the "implicit" category. In sum, people rely heavily on identifying not only visual information such as objects or events but also verbal information such as overlay text or letters on objects, when making sense of images (or videos).

Kim (2012) investigated how useful is speech summarization in speech form to that in text form. In order to compare summaries in speech form to those in text form, the author constructed a pilot system for text and spoken surrogates for 10 sample speeches; text surrogates in Korean were obtained from the Ted Talk site and then the text surrogates were "spoken" by a text-to-speech synthesizer to create the spoken surrogates. Next, the author asked each of recruited 46 participants, all of whom were undergraduate students pursuing majors in library and information science, to write the advantages and disadvantages of the spoken and text surrogates after watching both surrogates for each of the 10 videos.

The analysis results are as follows: Eighteen participants (39.1%) said that it was comfortable listening to the spoken surrogates; fifteen (32.6%) said that they focused well on the spoken ones; eight (17.4%) said that the spoken surrogates were played steadily and rapidly; six (13%) said that multitasking was possible when listening to the spoken ones (e.g., they can see visual images and listen to spoken surrogates at the same time); and four (8.7%) said that the spoken surrogates can preserve all acoustic/prosodic information such as pitch and duration, which leads to a better understanding of a video.

On the contrary, twenty (43.5%) said that in the spoken surrogates, it is hard to go to the exact part they want to replay; eight (17.4%) said that it became difficult to focus while listening a long spoken surrogate; and six (13%) said that the spoken surrogates are ephemeral and thus, they could not remember all of the content of the spoken surrogate.

Regarding the advantages and disadvantages of text surrogates, twenty–one (45.7%) said that it is very easy to go to the exact part that they want to read again; twelve (26.1%) said that text surrogates enabled them to gain an overall understanding of a video through the context of the sentences in the text surrogate; eleven (23.9%) said that text surrogates enabled them to control their speed of reading; and six (13%) said that it is easy to understand the meaning of unknown words (or phrases) because they were spelled out. On the contrary, sixteen (34.8%) said that it became difficult to focus or to read while reading a long text surrogate; seven (15.2%) said that reading text surrogates caused eye fatigue; and six (13%) said that considerably sized interfaces were required to browse text surrogates properly.

2.2.2 Multi-modal

Song and Marchionini (2007) compared the effectiveness of three different surrogates: visual alone (storyboard), audio alone (spoken description), and a combination of video and audio (a storyboard augmented with spoken description). The study showed that combined surrogates are more effective and, hence, strongly preferred. The authors also demonstrated that the use of only oral descriptions lead to better comprehension of the video segments than do only visual storyboards; however, people prefer to have visual surrogates and use them to confirm interpretations and add context.

Song, Marchionini, and Oh (2010) examined what features in the video catch the eyes and ears of human assessors in order to improve automatic video summaries. For a test, the authors created video summaries for instructional documentary videos and the video summaries were then examined by human assessors. They used six audio features (music, single human voice, multiple human voice, proper nouns, natural sound, and artificial sound) and eight visual features (text [superimposed names, locations], faces, graphics and logos, graphs, equations, animals, human built artifacts, and natural scenes) in order to index the video summaries.

Their study results showed that within the visual channel, text, equations, and graphs are important if audio is not available, otherwise, human faces and natural scenes are often selected for video summaries. On the other hand, within audio channels, single human voices and natural sounds are selected but multiple human voices are not as commonly selected for video summaries. Their result is in line with the study of Kim and Kim (2010) showing that participants rely heavily on identifying verbal information such as overlay text or letters on objects, when making sense of images (or videos) during viewing video storyboards that are composed of only key frames (images) without audio information.

Kim (2011) examined the interactive effect of spoken words and imagery not synchronized in audio/image surrogates for video gisting. To do that, the author conducted an experiment with 64 partici-pants, under the assumption that participants would better under-stand the content of videos when viewing audio/image surrogates rather than audio or image surrogates. The results of the experiment showed that overall audio/image surrogates were better than audio or image surrogates for video gisting, although the unsynchronized multimedia surrogates made it difficult for some participants to pay attention to both audio and image when they were not synchronized.

2.3 Narrative Theory

Chatman (1975) analyzed the narration structures of films and novels. He suggested that each narrative has two parts: a story, the content or chain of events (actions, happenings), plus what may be called the existents (characters, setting); and a discourse, that is, the expression, or the means by which the content is communicated. Chatman mentioned that the story is the "what" in a narrative that is depicted, and the discourse is the "how." Leaving the role of the discourse of a story aside, his argument regarding the structure of a story was quite simple, that is, a story equals events plus characters plus setting. As shown below, Chatman's model conceptualizes narrative transmission as follows:

Actual author → [implied author → (narrator) → (narratee) → implied audience] → actual audience

His model illustrates the flow of a narrative as originating from an actual author, written according to the convictions of the implied author, which is voiced and directed by a narrator to the narratee, who attempts to create an implied audience in the life and situation of a specific actual audience who physically hears or reads the narrative (De Milander, 2015). In other words, Chatman's model describes the basic concepts that need to be represented in some formal way so that a story engine can build a coherent narrative regarding content and presentation form. Thus, Chatman's model can be adapted for automated story generation (Reijnders, 2011).

Bordwell (1985) argued that there is no point of positing

communication model, which Chatman had set up as a consensus within narrative studies for studying the fundamental process to understand narrative movies. Chatman and Bordwell conflicted each other over several issues: whether the communication process is one- or two-sided; and what an implied author means in the context of narrative movies. The weakest point of Chatman's narrative transmission model is that no narrative can be transmitted by itself. Instead, the audience should reconstruct the narrative in the movie with the help of piecemeal information in the text.

Kim (2009) proposed a modified model of narrative communication introducing a notion of the space of communication, in front of which actual authors and audience posit themselves (see Figure 2-2). In his model, implied authors and implied audience by Chatman are nothing but psychological entities in the communication space that is mediated by audiovisual (AV) technologies. Producers should consider what audiences want to watch and hear, whereas audiences should consider what the producers would like their works to convey.

Applying this model of mediated narrative communication to Hichcock's classic movie "Rear Window," he tried to show the validity of the narrative communication model. He discussed the implications of the notion of the "space of communication" which is mediated technically by the camera in front of the director and the screen before the actual audience. The space of communication is found to relate the story-space of the movie and the space of the actual author and audience, removing the trace of mediation, thus creating non-mediation.

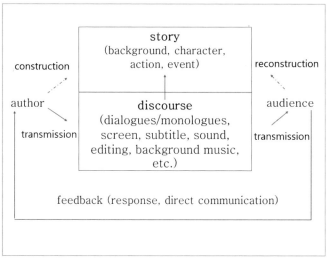

Figure 2-2. Modified model of narrative communication (Kim, 2009)

According to the model as shown in Figure 2-2, a sender (e.g., video producers) intends to deliver a message, and its meaning is reinterpreted and reconstructed from a receiver's point of view during the process of communication. Thus, understanding the narrative structure of a video helps to construct an efficient method for abstracting visual information (e.g., moving images). For example, in order to enable receivers (users) to better understand the structural analysis of a video, we need to obtain information on the story and the discourse of a narrative. That is, to construct an efficient video trailer, we need to extract shots from a video that includes topic–related events, characters, and background. Additionally, in order to obtain information on discursive features, we need the shots that engage in narrative activities, such as dialogues and monologues of characters, commentary, camera work and movement, and music.

References

Allegretti, M., Moshfeghi, Y., Hadjigeorgieva, M., Pollick, F. E., Jose, J. M., & Pasi, G. (2015). When relevance judgement is happening?: An EEG–based Study. Proceedings of the 38th International ACM SIGIR Conference on Research and Development in Information Retrieval (pp. 719~722). New York: ACM.

Bordwell, D. (1985). Narration in the fiction film. Madison, WI.: University of Wisconsin Press

Chatman, S. (1975). Towards a theory of narrative. New Literary History, 6(2), 295~318.

Chen, Z., & Xu, Y. (2005). User–oriented relevance judgment: A conceptual model. Proceedings of the 38th Annual Hawaii International Conference on System Sciences (pp. 101b~101b).

Choi, Y., & Rasmussen, E. M. (2002). Users' relevance criteria in image retrieval in American history. Information Processing and Management, 38(5), 695~726.

De Milander, C. (2015). Contemporary implications of the first—century counter—ethos of Jesus to the scripted universe of gender and health in John 4 & 9: A narrative—critical analysis (Doctoral dissertation, Stellenbosch: Stellenbosch University).

Ding, W., Marchionini, G., & Soergel, D. (1999). Multimodal surrogates for video browsing. Proceedings of the Fourth ACM Conference on Digital Libraries (pp. 85~93). New York: ACM Press.

Eakins, J. P., & Graham, M. E. (1999). Content—based image retrieval (JISC Technology Applications Programme Report 39). Retrieved January 31, 2010, from http://www.jisc.ac.uk/media/documents/programmes/jtap/ jtap—039.pdf.

Fitzgerald, M.A., & Galloway, C. (2001). Relevance judging, evaluation, and decision making in virtual libraries: A descriptive study. Journal of the American Society for Information Science and Technology, 52, 989~ 1010.

Greisdorf, H., & O'Connor, B. (2002). Modelling what users see when they look at images: A cognitive viewpoint. Journal of Documentation, 58 (1), 6~29.

Gwizdka, J., Hosseini, R., Cole, M., & Wang, S. (2017). Temporal dynamics of eye-tracking and EEG during reading and relevance decisions. Journal of the Association for Information Science and Technology, 68(10), 2299~2312.

Harter, S. P. (1992). Psychological relevance and information science. Journal of the American Society for Information Science and Technology, 43(9), 602~615.

Hendahewa, C., & Shah, C. (2015). Implicit search feature based approach to assist users in exploratory search tasks. Information Processing and Management, 51(5), 643~661.

Hendahewa, C., & Shah, C. (2017). Evaluating user search trails in exploratory search tasks. Information Processing and Management, 53(4), 905~922.

Huang, X., & Soergel, D. (2006). An evidence perspective on topical relevance types and its implications for exploratory and task-based retrieval. Information Research, 12(1), 12-1.

Huang, X. , & Soergel, D. (2013). Relevance: An improved framework for explicating the notion. Journal of the Association for Information Science and Technology, 64 (1), 18~35.

Hughes, A.,Wilkens, T.,Wildemuth, B.M., & Marchionini, G. (2003). Text or pictures? An eyetracking study of how people view digital video surrogates. In E.M. Bakker, T.S. Huang, M.S. Lew, N.Sebe, & X.(S.) Zhou (Eds.), Lecture Notes In Computer Science: Vol. 2728. Proceedings of the Conference on Image and Video Retrieval (CIVR), (pp. 271~280). Berlin, Germany: Springer.

Ingwersen, P. (1996). Cognitive perspectives of information retrieval interaction. Journal of Documentation, 52(1), 3~50.

Kim, H. (2008). Design and evaluation of the key-frame extraction algorithm for constructing the virtual storyboard surrogates. Journal of Korean

Society for Information Management, 25(4), 131~148.

Kim, H. (2011). A study on the interactive effect of spoken words and imagery not synchronized in multimedia surrogates for video gisting. Journal of the Korean Society for Library and Information Science, 45(2), 97~118.

Kim, H. (2012). A Tag-based framework for extracting spoken surrogates. Proceedings of the ASIST Annual Meeting, 49. Medford, NJ: Information Today.

Kim, H., & Kim, Y. (2010). Toward a conceptual framework of key-frame extraction and storyboard display for video summarization. Journal of the American Society for Information Science and Technology, 61(5), 927~939.

Kim, Y. (2009). A structural model of mediated visual communication in narrative movies: Focusing on Chatman and Bordwell's controversy. Korean Journal of Journalism and Communication Studies, 53(1), 209~232.

Komaki, D., Hara, T., & Nishio, S. (2012). How does mobile context affect people's web search behavior?: A diary study of mobile information needs and search behaviors. In Advanced Information Networking and Applications (AINA), 2012 IEEE 26th International Conference on (pp. 245~252). IEEE.

Maglaughlin, K. L., & Sonnenwald, D. H. (2002). User perspectives on relevance criteria: A comparison among relevant, partially relevant, and not-relevant judgments. Journal of the American Society for Information Science and Technology, 53(5), 327~342.

Moshfeghi, Y., Pinto, L. R., Pollick, F. E., & Jose, J. M. (2013). Understanding relevance: An fMRI study. In P. Serdyukov et al., eds. European Conference on Information Retrieval (pp. 14~25). Springer Berlin Heidelberg.

Mostafa, J., & Gwizdka, J. (2016). Deepening the role of the user: Neuro-physiological evidence as a basis for studying and improving search. Proceedings of the 2016 ACM on Conference on Human Information Interaction and Retrieval (pp. 63~70). New York: ACM.

Panofsky, E. (1955). Meaning in the visual arts: Meaning in and on art history. Garden City, NY: Doubleday.

Paivio, A. (1990). Mental representations: A dual coding approach. Oxford University Press.

Reijnders, K. (2011). Suspense Tours: Narrative generation in the context of tourism. Thesis Master Information Studies Programme of Human Centered Multimedia, Universiteit van Amsterdam.

Rieh, S. Y., Collins-Thompson, K., Hansen, P., & Lee, H. J. (2016). Towards searching as a learning process: A review of current perspectives and future directions. Journal of Information Science, 42(1), 19~34.

Ruthven, I. (2014). Relevance behaviour in TREC. Journal of Documentation, 70, 1098~1117.

Saracevic, T. (2007). Relevance: A review of the literature and a framework for thinking on the notion in information science. Part II: Nature and manifestations of relevance. Journal of the Association for Information

Science and Technology, 58(13), 1915~1933.

Shatford, S. (1986). Analyzing the subject of a picture: A theoretical approach. Cataloging & Classification Quarterly, 6(3): 39~62.

Singer, G., Danilov, D., & Norbisrath, U. (2012). Complex search: Aggregation, discovery, and synthesis. Proceedings of the Estonian Academy of Sciences, 61, 89~106.

Singer, G., Norbisrath, U., & Lewandowski, D. (2013). Ordinary search engine users carrying out complex search tasks. Journal of Information Science, 39(3), 346~358.

Song, Y., & Marchionini, G. (2007). Effects of audio and visual surrogates for making sense of digital video. Proceedings of CHI 2007 (pp. 867~876), San Jose, CA, USA.

Song, Y., Marchionini, G., & Oh, C. Y. (2010). What are the most eye-catching and ear-catching features in the video?: Implications for video summarization. In Proceedings of the 19th international conference on World wide web (pp. 911~920). ACM.

Swanson, D. R. (1986). Subjective versus objective relevance in bibliographic retrieval systems. The Library Quarterly, 56(4), 389~398.

Taylor, A. (2012). User relevance criteria choices and the information search process. Information Processing and Management, 48, 136~153.

Turner, J. (1994). Determining the subject content of still and moving documents for storage and retrieval: An experimental investigation. Unpublished Ph.D. Dissertation. University of Toronto.

Wildemuth, B., Marchionini, G., Wilkens, T., Yang, M., Geisler, G., Fowler, B., ... & Mu, X. (2002). Alternative surrogates for video objects in a digital library: Users' perspectives on their relative usability. Research and Advanced Technology for Digital Libraries, 143~160.

Wilson, P. (1973). Situational relevance. Information Storage and Retrieval, 9(8), 457~471.

Yang, M., & Marchionini, G. (2004). Exploring users' video relevance criteria: A pilot study. Proceedings of the ASIST Annual Meeting, 41 (pp. 229~238). Medford, NJ: Information Today.

Chapter 3

Research Method:
Focusing on NP Measures

3. Research Method: Focusing on NP Measures

Although a self-reporting method (e.g., survey, questionnaire, or poll) is frequently used in many studies and has been found to have an acceptable validity and reliability, this method has been a frequent target of criticism because it does not identify the immediate reactions of respondents. Furthermore, it has the limitation that it cannot avoid the errors caused by the respondents when they express their thoughts.

Thus, to obtain objective and scientific research results, NP methods can be utilized to measure the cognitive and emotional reactions of humans, thus eliminating the limitations occurring due to the nature of self-reporting measurements. Furthermore, research results obtained through NP methods can be utilized to design a brain-computer interface (BCI).

In this chapter, we introduce NP studies on cognitive informatics to describe and explain the generation of NP evidence using three types of modalities, namely eye tracking, EEG, and fMRI. The case study titled, "Understanding topical relevance judgment in visual simple search: An EEG study" written by the author is also described.

3.1 EEG

EEG is a method of measuring the electrical activity of the brain by recording the electrical current captured on the scalp, under the assumption that a human visual system responds differently (Lawhern, Slayback, Wu, & Lance, 2015). Depending on the type of data, its response result is reflected in people's brain waves. The standard or common EEG wave patterns include theta, alpha, beta, and gamma waves. The common wave patterns represent specific frequency and amplitude patterns. Based on alpha, beta, theta, and gamma waves, we can capture and analyze brain signals that do not use special stimuli, drawing general conclusions regarding the status of the participant's functioning brain. It is known that when a person is in a relaxed mood and/or has his or her eyes closed, alpha waves are particularly prominent, and in contrast, when a person is in an anxious state, beta waves appear predominantly (Figure 3−1).

In contrast to investigating the current status of a brain, there are responses of the brain to specific stimuli. These responses to the stimuli are termed event−related potentials (ERPs) (Makeig, 2009; Mostafa & Gwizdka, 2016). For example, P3b, which has a positive peak around 400 ms, might occur when a stimulus is a rare or relevant target (Fernandes, Ferreira, Almeida, & Dias, 2015), whereas N400, which has a negative peak around 400 ms, might occur when a stimulus does not correspond to a prior topical expectation (Vō & Wolfe, 2013). Additionally, P600 is known to be related to topic shift and discourse–internal reorganization and integration (Xu & Zhou, 2016; Wang & Schumacher, 2013; Burmester, Spalek, & Wartenburger, 2014).

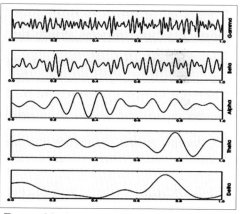

Figure 3-1. Types of brain waves (https://www.quantumneurocare.com)

EEG has been used in many studies, such as relevance judgment (Allegretti, et al., 2015; González-Ibáñez, Escobar-Macaya, & Manriquez, 2016), visual search (Brouwer, Hogervorst, Oudejans, Ries, & Touryan, 2017), video summarization (Mehmood et al.,

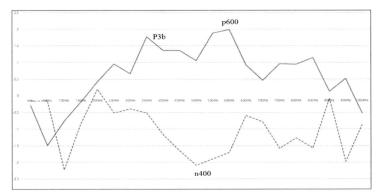

Figure 3-2. Grand-average ERP waveforms elicited at electrode Cz associated with irrelevant stimuli (dashed) and relevant stimuli (solid).

2016), and tagging (Koelstra, Muehl, & Patras, 2009). Koelstra et al. (2009) proposed a method for automatically assigning implicit tags to a video. They conducted an ERP experiment wherein 49 videos were utilized as stimuli. During the experiment, each of 17 subjects was asked to decide whether or not the topic of a given tag matched the topic of a video. The authors found lower minimum potentials in brain wave responses to topic—irrelevant tags than to topic—relevant tags in the N400 time window.

Allegretti, et al. (2015) conducted an ERP—based study on the first 800 ms of a relevance assessment process to determine the time at which topic relevance is assessed in the brain. Twenty participants were asked to provide an explicit judgment regarding the relevance of a presented image according to a given topic. The authors found greater maximum potentials for brain wave responses to topic–relevant images than to topic–irrelevant images in the frontal area in the 180~300–ms time window, and in the central and central–parietal areas in the 300~500–ms time window. They also reported that the central area had the greatest difference between relevant and irrelevant images in the 500~800–ms time window compared to the other time windows. Together, these results suggest that relevance judgments for visual stimuli are made at a time window around 800 ms.

Mehmood et al. (2016) proposed a human–attention model that combines both multimedia content and a viewer's neuronal responses for video summarization. In their model, neuronal attention is computed using beta–band EEG frequencies of neuronal responses under the assumption that the power of beta–band activity is related to the level of human attention (Gola, Kamiński, Brzezicka, & Wróbel, 2012).

3.2 fMRI

fMRI is an imaging technique using MRI technology that measures brain activity by detecting changes associated with blood flow (Huettel, Song, & McCarthy, 2014). When conducting a specific task, oxygenated blood flows toward the specialized brain areas relevant to those tasks, and is measured by the blood–oxygenation level dependent measure (BOLD) (Alpert, Badgaiyan, Livni, & Fischman, 2003).

In contrast to an EEG, the mapping of active areas based on BOLD is highly precise, and it is typically captured as voxels in the range of millimeters (Mostafa & Gwizdka, 2016). However, fMRI is not suitable for tracking cerebral dynamics within milliseconds.

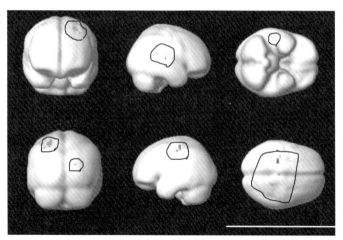

Figure 3-3. Brain-surface area mapping of fMRI BOLD signals collected from 20 subjects who use digital learning contents (Kwon et al., 2015)

fMRI has been used in studies, such as relevance judgment and requirements for information (Moshfeghi et al., 2013; Moshfeghi, Triantafillou, & Pollick, 2016). For example, Moshfeghi et al.

(2013) conducted an fMRI study in order to investigate the connection between relevance and brain activity. For this purpose, the brain activity of 18 subjects was measured while performing their four topical relevance assessment tasks on relevant and irrelevant images. The four topics were: "Olympic Torch," "Fortis Logo," "Obama Clinton," and "British Royals." During the test, each subject was asked to press button 1 (if the image was relevant) or 2 (if the image was irrelevant) after being asked to view an image.

They found that fronto–parietal circuits are more activated by topic–relevant images than by topic–irrelevant images. However, they questioned whether fronto–parietal activity reflects specific aspects of relevance or whether it simply reflects greater attention to topic–relevant images than topic–irrelevant images, as fronto–parietal circuits are also known to be activated by attention to stimuli (Corbetta & Shulman, 2002).

3.3 Eye Tracking

Eye tracking has a long history in psychological and medical research as a tool for recording and studying human visual behavior (Majaranta & Bulling, 2014). Eye tracking refers to the process of tracking eye movements or the absolute point of a gaze–referring to the point a user's gaze is focused on in a visual scene (Yousefi, Karan, Mohammadpour, & Asadi, 2015; Majaranta & Bulling, 2014).

Eye tracking is useful in a broad range of application areas, from psychological research and information science studies to usability studies and interactive, gaze—controlled applications. An eye tracker is a portable hardware device that performs this measurement by measuring eye movement (Murray et al., 2009).

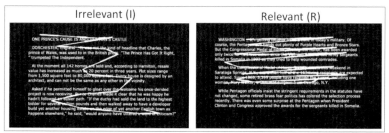

Figure 3-4. Eve movement reading patterns on irrelevant and relevant document (Gwizdka, 2014).

Eye tracking has been used in information science studies, such as relevance judgment (Gwizdka, 2014; Gwizdka, 2017), information retrieval behavior (Bilal & Gwizdka, 2016; Gwizdka & Zhang, 2015; Brouwer et al., 2017), and human—computer interaction (Sharma & Dubey, 2014).

Gwizdka (2014) investigated the differences in reading patterns and in the cognitive effort between text documents of different degrees of relevance. He found that the degree of relevance of a text document does affect how it is read and that it does also affect the level of cognitive effort required to read it. In addition, significant differences in reading patterns were found between documents at the three levels of relevance (relevant, partially relevant, and irrelevant). As shown in Figure 3–4, his study result shows that relevant documents tended to be read more coherently continuously, whereas irrelevant documents tended to be scanned.

Brouwer et al. (2017) examined the differences between verbal–izers and visualizers in the way they gaze at texts and pictures during learning. Using questionnaires, university students were classified according to their verbal or visual cognitive style, and they were asked to learn about two different topics by means of text–picture combinations. Eye tracking was used to investigate their gaze behavior. Their results showed that verbalizers spent significantly more time inspecting texts than visualizers, whereas visualizers spent more time inspecting pictures than verbalizers, supporting not only the existence of the visual–verbal cognitive style but also its influence on learning behavior.

Gwizdka, Hosseini, Cole, and Wang (2017) proposed a method to investigate processes of reading and subjective relevance judgments in search using eye tracking and EEG. The authors found differences in cognitive processes used to assess texts of varied relevance levels and evidence for the potential to detect these differences in search sessions.

3.4 Understanding Topical Relevance Judgment in Visual Simple Search: An EEG Study

3.4.1 Introduction

This study focuses only on topical relevance judgments during a simple search, wherein a search topic is defined before a search. To understand how topical relevance judgments during a simple search for visual information occur in the brain, we used event−related potential (ERP) components that are used to measure users' cognitive responses evoked by visual stimuli.

West and Holcomb (2002) found that the ERPs elicited by a congruous final picture had a small negativity (N300) and that a large extended P300 occurs following N300. The authors also found that the ERPs elicited by an incongruous final picture had two separate negative components, N300 and N400, which were followed by a long slow P300.

As such, ERP studies on images have occasionally found N300, which may have been elicited due to its involvement in the semantic processing of images with no linguistic mediation or interference (West & Holcomb, 2002). The N300/N400 effects for semantic violations were also observed for object–scene incongruity (Mudrik et al., 2014; Võ & Wolfe, 2013). We speculate that a fronto–central N300, which is related to categorical mismatch (Hamm, Johnson, & Kirk, 2002), is suitable for semantic categorization, and that relevance judgment occurs following semantic categorization.

3.4.2 Hypotheses

Our hypotheses for a simple search are as follows:

Hypothesis 1 (a fronto–central N300 effect hypothesis): the amplitude of the negative peak around 300 ms, occurring in frontal and central areas in case of topic–irrelevant stimuli, will be much lower than that in case of topic–relevant stimuli.

Hypothesis 2 (a posterior P3b effect hypothesis): the amplitude of the positive peak around 400 ms, occurring in a posterior area in case of topic–relevant stimuli, will be much higher than that in case of topic–irrelevant stimuli.

3.4.3 Experiments

1) Participants and Sample Videos

We conducted an experiment using a within−subject design wherein the independent variable was relevance. This variable had one of three values: relevant, partially relevant, and irrelevant. The dependent variables were EEG signal values. This research was approved by the Institutional Review Board of Myongji University. Written informed consent was obtained from each participant prior to his or her involvement in the experimental sessions.

We decided to use videos in the same genre (documentary videos) as experimental materials because the characteristics of videos seem to vary according to genre. We selected three videos from YouTube and Korea Heritage (http://www.k−heritage.tv/) based on the following criteria: short videos whose duration was between 1 min and 3 min, and videos focusing on one topic. The three videos lasted from 1 min and 9 s to 2 min and 11 s, and consisted of 9 to 12 scenes each. The experimental materials used for the experiment are shown in Table 3−1.

In our experiment, we also utilized twenty−seven shots extracted from the three videos as stimuli. Two coders evaluated all shots belonging to each video and selected the three most relevant, three partially relevant, and three irrelevant shots from each video based on the degree of topic relevance. This resulted in the selection of nine shots per video.

We recruited 25 participants (23 undergraduate and 2 graduate students) from Myongji University by email and phone. The participants were compensated with money. The participants were limited to right–handed male students in the age range of 21~27 years to minimize differences between individual EEGs.

Table 3-1. Details of the sample videos

No.	Title	Description	Genre
1	Discovering glasses	This video explains how glasses began to be used and for what purposes they are used.	Documentary (education)
2	Uhm's bicycle	Bicycle history in Korea was described by introducing Uhm, a Korean famous cyclist.	Documentary (history)
3	Two pocket watches	The video describes the story about two watches.	Documentary (history)

2) Procedure

The experiments were carried out in the following manner. The participants were provided with instructions for the experiment. We asked the participants to focus on the topic of the video and to memorize shots reflective of the central theme of the video while watching the video. Figure 3–5 shows how the videos and their shots were presented to the participants. We used the three videos and their twenty–seven shots extracted from them (nine shots per video) as stimuli. At the beginning, one stimulus, which was randomly chosen from the three sample videos, was shown to the participants. Before the presentation of each video, a 3−s black screen with a fixation cross was presented and then a 500−ms black screen was presented.

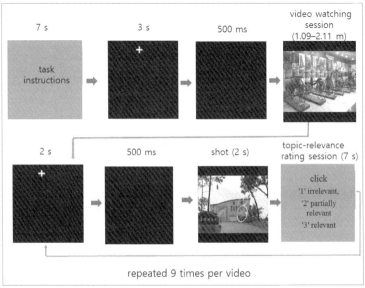

Figure 3-5. Order and timing of the experiment

After being asked to watch a video, each of the nine shots in the video was sequentially presented for 2 s to the participants. Before the presentation of each shot, a 2−s black screen with a fixation cross was presented and then a 500−ms black screen was presented. After the presentation of each shot, the participants were asked to rate each of the nine shots belonged to the video by pressing the "1" key on the computer keyboard if the presented shot was irrelevant to the topic of the video, the "2" key on the computer keyboard if it was partially relevant, or the "3" key on the computer keyboard if it was relevant. The participants were required to respond within 7 s. The above steps were repeated for the other two videos.

3.4.4 EEG Recording and Processing

1) EEG Recording

A Neuroscan Synamp amplifier (Compumedics Neuroscan; Victoria, Australia) linked to a Quick Cap with 32 channels was connected to two computers, one of which handled brainwave collection and data storing, and the other computer handled the display and E-Prime presentation. Electrodes on the Quick Cap include the FP1/FP2, F3/Fz/F4, F7/F8, FC3/FCz/FC4, CP3/CPz/CP4, C3/Cz/C4, P7/P8, P3/Pz/P4, O1/Oz/O2, T7/T8, TP7/TP8, and FT7/FT8. Data were digitally sampled for analysis at a 1,000-Hz sampling rate. EEG recordings were made with reference to the electrically linked mastoids A1 and A2. Eye movements and other artifacts were monitored using 2 electrodes (vertical and horizontal electro-oculograms), while the other 30 channels were placed on the scalp at locations based on the International 10–20 system.

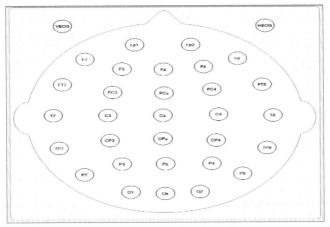

Figure 3-6. International 10-20 electrode montage

2) Data Analysis

EEG data were analyzed using a 7.09 CURRY program (Compu‑medics Neuroscan). Noise due to artifacts such as eye movements was removed using a covariance matrix. EEG data from 4 out of the 25 participants were eliminated from analysis due to excessive noise in the responses, or failure to follow task instructions. The mean values from each participant's EEG data during the 200 ms before stimulus onset were adjusted to zero as a baseline correction for the participant's background brainwave. ERP brainwave data were collected during the 1,000 ms after stimulus onset. Thus, EEG data were analyzed mainly over a time period of 1,200 ms (−200 ms to 1,000 ms).

EEG signals obtained during the topic−relevance rating session were preprocessed using the 7.09 CURRY program. The preprocessing steps consisted of setting a band−pass filter from 0.1 to 35 Hz, which was applied to remove power−line noise. Epochs were then extracted from 200 ms before stimulus presentation to 1,000 ms after stimulus presentation. Among the 567 epochs (27 epochs per participant) obtained from the 21 participants, 181 epochs were found to be topic−relevant, 203 epochs were partially relevant, and 183 epochs were topic−irrelevant.

3.4.5 Results

We performed a repeated−measures t−test with a within−subject factor of topic−relevance (1: topic−irrelevant, 3: topic−relevant) to verify the two hypotheses. We used only topic−relevant and topic−irrelevant shots. Partially relevant shots were excluded due to their vague characteristics.

1) ERP Results

The grand−average ERP waveforms elicited at electrodes Fz, Cz, and Pz are shown in Figure 3−7. As shown in the upper and middle of Figure 3-7, each of the Fz channels for the two cases (relevant and irrelevant) shows a clear occurrence of N300 after an N100 occurrence, whereas each of the Cz channels shows the same ERP component occurrences as in the Fz channels.

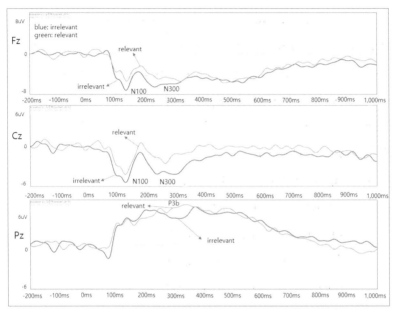

Figure 3-7. Grand-average ERP waveforms elicited at electrodes Fz, Cz, and Pz associated with irrelevant shots and relevant shots (among the 567 epochs obtained from the 21 participants, 181 epochs found to be topic-relevant and 183 epochs to be topic-irrelevant were analyzed for the grand-average ERP waveforms).

The grand-average ERP waveform at the Pz electrode elicited by irrelevant shots had a positivity that peaked at ~380 ms (P3b). In contrast, relevant shots led to a positivity that peaked at ~350 ms (P3b). The positivity is similar to that was observed in response to irrelevant shots, but has larger amplitude.

2) Results of Hypothesis Testing

We performed a repeated−measures t−test with a within−subject factor of topic−relevance (1: topic−irrelevant, 3: topic−relevant) to verify the two hypotheses. We applied a Bonferroni correction, which is an adjustment made to the alpha value when several statistical tests are being performed simultaneously on a single dataset (Abdi, 2007). We performed 30 comparisons in the current analysis. Therefore, the critical p value used to determine statistical significance was 0.05 (the alpha level) divided by 30, or 0.0017.

(1) Hypothesis 1: We examined whether a fronto−central N300 effect appears for video shots so that a more negative ERP signal is observed for topic−irrelevant shots than for topic−relevant shots using a repeated measure t−test with a within−subject factor of topic−relevance (1: topic−irrelevant, 3: topic−relevant). At the statistical significant level of 0.05, we found the lower minimum potentials of brain wave responses to topic−irrelevant shots than to topic−relevant shots at the left central region (C3), the right and left

central-parietal regions (CP3 & CP4), and the midline central-parietal region (CPz) (Table 3-2). Thus the hypo-thesis 1 was accepted. However, none of these values remained significant after the Bonferroni correction (p < 0.0017).

Table 3-2. Means of minimums at 250~350ms (N300) between topic-relevant and topic-irrelevant video shots (unit: μV)

CH	Mean (SD)		t (p)	CH	Mean (SD)		t (p)
	Irrelevant	Relevant			Irrelevant	Relevant	
C3	−6.42 (4.83)	−3.90 (4.21)	4.78 (.049)*	CPz	−4.47 (4.32)	−1.81 (4.26)	6.83 (.023)*
CP3	−3.14 (4.50)	−0.77 (3.83)	5.33 (.040)*	CP4	−3.61 (3.24)	−0.97 (2.47)	7.43 (.018)*

*p < 0.05, **p < 0.01, ***p < 0.0017 (after Bonferroni correction), SD: standard deviation

(2) Hypothesis 2: It is hypothesized that the posterior P3b effect appears both for the topic-relevant and topic-irrelevant shots, eliciting a more positive ERP signal in response to the topic-relevant shots than to the topic-irrelevant shots. The test result shows a clear occurrence of p3b and a significant difference in p3b activation between the two cases (Table 3-3).

At the statistical significant level of 0.01, we found the higher maximum potentials of brain wave responses to topic—relevant shots than to topic—irrelevant shots at the left parietal lobes (P3 & P7), the left temporal—parietal region (TP7), the left and right central regions including the midline central region (C3, Cz, & C4), the left frontal—central region (FC3), and the left central—parietal region (CP3).

Table 3-3. Means of maximums at 350~500 ms (P3b) between topic-relevant and topic-irrelevant shots (unit: μV)

CH	Mean (SD)		t (p)	CH	Mean (SD)		t (p)
	Irrelevant	Relevant			Irrelevant	Relevant	
P3	7.95 (4.36)	10.67 (4.67)	13.98 (.003)**	Cz	2.43 (4.94)	5.21 (6.16)	11.72 (.005)**
Pz	6.64 (4.91)	9.13 (4.98)	6.02 (.030)*	C4	2.11 (3.90)	4.34 (5.07)	9.54 (.009)**
P7	7.10 (3.29)	9.11 (3.02)	9.50 (.009)**	FT7	−0.79 (4.36)	1.66 (4.95)	6.92 (.022)*
T8	0.10 (2.22)	1.69 (3.48)	4.85 (.048)*	FC3	−0.09 (4.89)	2.99 (5.25)	10.97 (.006)**
TP7	3.27 (3.41)	5.91 (3.38)	13.58 (.003)**	FCz	0.07 (5.12)	2.16 (6.07)	6.69 (.024)*
Fz	−0.44 (4.48)	1.26 (5.24)	9.24 (.010)*	FC4	−0.10 (3.40)	1.67 (5.01)	5.77 (.033)*
F4	−1.29 (4.05)	0.34 (4.63)	7.72 (.017)*	CP3	4.75 (5.18)	7.68 (5.00)	13.39 (.005)**
C3	1.78 (4.90)	5.19 (5.33)	22.48 (.000)***	CPz	4.53 (5.38)	7.19 (5.54)	8.67 (.012)*

*p < 0.05, **p < 0.01, ***p < 0.0017 (after Bonferroni correction), SD: standard deviation

In addition, at the statistical significance level of 0.05, the more positive potentials for topic—relevant when compared to topic—irrelevant shots were elicited at the midline parietal region (Pz), the right temporal lobe (T8), the midline and right frontal regions (Fz & F4), the left frontal–temporal region (FT7), the midline and right frontal–central regions (FCz & FC4), and the midline central–parietal lobe (CPz). Thus, the hypothesis 2 was accepted. The difference in responses at the C3 remained significant after the Bonferroni correction (p < 0.0017).

References

Abdi, H. (2007). Bonferroni and Šidák corrections for multiple comparisons. Encyclopedia of measurement and statistics, 3, 103~107.

Allegretti, M., Moshfeghi, Y., Hadjigeorgieva, M., Pollick, F. E., Jose, J. M., & Pasi, G. (2015). When relevance judgement is happening?: An EEG–based Study. Proceedings of the 38th International ACM SIGIR Conference on Research and Development in Information Retrieval (pp. 719~722). New York: ACM. (Chapter 2).

Alpert, N. M., Badgaiyan, R. D., Livni, E., & Fischman, A. J. (2003). A novel method for noninvasive detection of neuromodulatory changes in specific neurotransmitter systems. Neuroimage, 19(3), 1049~1060.

Bilal, D., & Gwizdka, J. (2016). Children's eye-fixations on google search results. Proceedings of the Association for Information Science and Technology, 53(1), 1~6.

Brouwer, A. M., Hogervorst, M. A., Oudejans, B., Ries, A. J., & Touryan, J.

(2017). EEG and eye tracking signatures of target encoding during structured visual search. Frontiers in Human Neuroscience, 11, 264.

Burmester, J., Spalek, K., & Wartenburger, I. (2014). Context updating during sentence comprehension: The effect of aboutness topic. Brain and Language, 137, 62~76.

Corbetta, M., & Shulman, G. (2002). Control of goal-directed and stimulus-driven attention in the brain. Nature Reviews Neuroscience, 3(3), 201~215.

Fernandes, L. S., Ferreira, D. S., Almeida, P. R., & Dias, N. S. (2015). Aging and attentional set shifting on WCST: An event-related EEG study. In Neural Engineering (NER), 2015 7th International IEEE/EMBS Conference on (pp. 1088~1091). IEEE.

Gola, M., Kamiński, J., Brzezicka, A., & Wróbel, A. (2012). Beta band oscillations as a correlate of alertness-changes in aging. International Journal of Psychophysiology, 85(1), 62~67.

González-Ibáñez, R., Escobar-Macaya, M., & Manriquez, M. (2016). Using low-cost electroencephalography (EEG) sensor to identify perceived relevance on web search. Proceedings of the Association for Information Science and Technology, 53(1), 1~5.

Gwizdka, J. (2014). Characterizing relevance with eye-tracking measures. In Proceedings of the 5th Information Interaction in Context Symposium (pp. 58~67). ACM.

Gwizdka, J. (2017). Differences in reading between word search and information relevance decisions: Evidence from eye-tracking. In Information Systems and Neuroscience (pp. 141~147). Springer

International Publishing.

Gwizdka, J., Hosseini, R., Cole, M., & Wang, S. (2017). Temporal dynamics of eye-tracking and EEG during reading and relevance decisions. Journal of the Association for Information Science and Technology, 68(10), 2299~2312.

Gwizdka, J., & Zhang, Y. (2015). Differences in eye-tracking measures between visits and revisits to relevant and irrelevant Web pages. In Proceedings of the 38th International ACM SIGIR Conference on Research and Development in Information Retrieval (pp. 811~814). ACM.

Hamm, J. P., Johnson, B. W., & Kirk, I. J. (2002). Comparison of the N300 and N400 ERPs to picture stimuli in congruent and incongruent contexts. Clinical Neurophysiology, 113(8), 1339~1350.

Huettel, S. A., Song, A. W, & McCarthy, G. (2014). Functional magnetic resonance imaging (3rd edition), Massachusetts: Sinauer Associates.

Koelstra, S., Mühl, C., & Patras, I. (2009). EEG analysis for implicit tagging of video data. In Affective Computing and Intelligent Interaction and Workshops, 2009. ACII 2009. 3rd International Conference on (pp. 1~6). IEEE.

Kwon, S., Kim, Y, Jung, D, Tae, M., & Kwon, Y. (2015). Brain activations during biology with digital learning contents of smartpad: fMRI study. School Science Journal, 9(2), 85~93.

Lawhern, V., Slayback, D., Wu, D., & Lance, B. J. (2015). Efficient labeling of EEG signal artifacts using active learning. In Systems, Man, and Cybernetics (SMC), 2015 IEEE International Conference on (pp.

3217~3222). IEEE.

Majaranta, P., & Bulling, A. (2014). Eye tracking and eye-based human-computer interaction. In Advances in Physiological Computing (pp. 39~65). Springer London.

Makeig, S. (2009). Electrophysiology: EEG and ERP analysis. Encyclopedia of Neuroscience. Academic Press. 879~882.

Mehmood, I., Sajjad, M., Rho, S., & Baik, S. W. (2016). Divide-and-conquer based summarization framework for extracting affective video content. Neurocomputing, 174, 393~403.

Moshfeghi, Y., Pinto, L. R., Pollick, F. E., & Jose, J. M. (2013). Understanding relevance: An fMRI study. In P. Serdyukov et al., eds. European Conference on Information Retrieval (pp. 14~25). Springer Berlin Heidelberg. (Chapter 2).

Moshfeghi, Y., Triantafillou, P., & Pollick, F. E. (2016). Understanding information need: An fMRI study. Proceedings of the 39th International ACM SIGIR Conference on Research and Development in Information Retrieval (pp. 335~344). ACM.

Mostafa, J., & Gwizdka, J. (2016). Deepening the role of the user: Neuro-physiological evidence as a basis for studying and improving search. Proceedings of the 2016 ACM on Conference on Human Information Interaction and Retrieval (pp. 63~70). New York: ACM.

Mudrik, L., Shalgi, S., Lamy, D., & Deouell, L. Y. (2014). Synchronous contextual irregularities affect early scene processing: Replication and extension. Neuropsychologia, 56, 447~458.

Murray, I. C., Fleck, B. W., Brash, H. M., MacRae, M. E., Tan, L. L., &

Minns, R. A. (2009). Feasibility of saccadic vector optokinetic perimetry: A method of automated static perimetry for children using eye tracking. Ophthalmology, 116(10), 2017~2026.

Sharma, C., & Dubey, S. K. (2014, March). Analysis of eye tracking techniques in usability and HCI perspective. In Computing for Sustainable Global Development (INDIACom), 2014 International Conference on (pp. 607~612). IEEE.

Võ, M. L. H., & Wolfe, J. M. (2013). Differential ERP signatures elicited by semantic and syntactic processing in scenes. Psychological Science, 24(9), 1816.

Wang, L., & Schumacher, P. B. (2013). New is not always costly: Evidence from online processing of topic and contrast in Japanese. Frontiers in Psychology, 4, 363.

West, W. C., & Holcomb, P. J. (2002). Event-related potentials during discourse-level semantic integration of complex pictures. Cognitive Brain Research, 13(3), 363~375.

Xu, X., & Zhou, X. (2016). Topic shift impairs pronoun resolution during sentence comprehension: Evidence from event-related potentials. Psychophysiology, 53(2), 129~142.

Yousefi, M. V., Karan, E., Mohammadpour, A., & Asadi, S. (2015). Implementing eye tracking technology in the construction process. Proceedings of 51st ASC Annual International Conference, College Station, TX.

Chapter 4

Cognitive Models for Multimedia Comprehension

4. Cognitive Models for Multimedia Comprehension

The process of multimedia comprehension can be described in two sequential phases: perception and integration (Mayer, 2005; Zhu, Goldberg, Eldawy, Dyer, & Strock, 2007). The perception process handles AV elements, whereas the integration process involves integrating the results of perceptual information processing with the preprocessed AV elements supporting a provisional topic. This leads to an assessment of the multimedia as either being a similar topic or a newly updated topic. This renewal or updating process of the topic by the viewer must be a complex cognitive process.

In this chapter, to examine the perception and integration processes precisely, we first reviewed Mayer's cognitive model of multimedia learning and Baddeley's WM theory (Mayer, 2005; Baddeley, 2007). Then, we constructed a relevance model wherein the concepts generated via the processing of newly integrated AV elements are integrated with existing concepts, with prior knowledge from long–term memory.

4.1 Baddeley's WM

WM refers to the system that is assumed to be necessary in order to keep things in mind while conducting complex tasks such as reasoning, comprehension, and learning (Baddeley, 2010). WM is a limited capacity memory system that holds information in temporary storage under attentional control when conducting a cognitive task. In its operation, WM combines percepts with memory.

WM has four types of components: a central executive (CE), two content-dedicated workspaces, a visuospatial sketchpad and phono-logical loop, and an episodic buffer (Baddeley, 2007). Two workspaces process information related to sound, vision, space, and

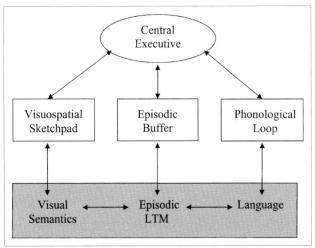

Figure 4-1. The working model of memory (Baddeley, 2010)

language, which are the most prominent areas of the human mind (Dudai, 2008). Below, we detail each of the four components of Baddeley's WM system (Figure 4–1).

The CE integrates information from the phonological loop, the visuospatial sketchpad, the episodic buffer, and long–term memory. The CE also plays an important role in focusing attention, selecting strategies, transforming information, suppressing irrelevant inform–ation, switching between tasks, and coordinating behavior (Baddeley, 2012; Baddeley, Eysenck, & Anderson, 2009; Reuter–Lorenz & Jonides, 2007). The CE seems to be connected to activation in the frontal lobe of the brain (Baddeley, 2006; Kolb & Whishaw, 2009).

A phonological loop deals with speech–based information. It comprises a phonological store and an articulatory rehearsal compo–nent where information is maintained by vocal or subvocal rehearsal and visually presented language can be transformed into phonological code by silent articulation, thereby encoding it into the phonological store.

The visio–spatial sketchpad enables a scene to be watched and the visual information about objects and characters to be gathered, and to navigate further from one location to another location (Logie, 2011). In other words, it deals with visuospatial information, which can be further fractionated into visuals or objects and spatial stores (Repov & Baddeley, 2006). It is suggested that the visio–spatial sketchpad activates different areas depending on the types of information such as visual, object and spatial information.

For example, Ventre–Dominey et al. (2005) suggested that spatial tasks stimulated activity in the parieto–occipital cortex, and the dorsal prefrontal cortex, whereas object tasks stimulated activity in the temporo–occipital cortex, the ventral prefrontal cortex, and the striatum.

Finally, the episodic buffer is portrayed as a temporary memory space wherein auditory, visual, and spatial information can all be combined with the information from long–term memory under the control of the CE. In other words, it integrates information from different modalities (Baddeley et al., 2009). The episodic buffer enables the creation of a complex representation of an event, which can be stored in long–term memory.

4.2 Mayer Model

Mayer's model is based on three assumptions regarding the manner in which the human brain works: dual channels, limited capacity, and active processing. The dual–channel assumption states that humans have two different systems for processing verbal and pictorial material. This is also a central feature of the theory of

Baddeley's WM.

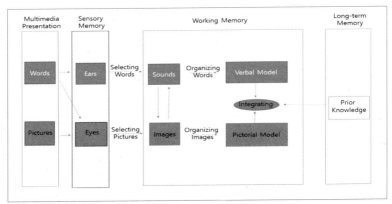

Figure 4-2. A cognitive model of multimedia learning by Mayer (2005)

The limited—capacity concept describes the human limitation for processing information in each channel at one time. Finally, the active processing assumption concerns the human ability to organize relevant information into coherent mental representations and to integrate these representations with prior knowledge from long—term memory (LTM) (Mayer, 2005).

We use Mayer's model to explain how viewers determine the topic of a video. Words and pictures in the form of a video presentation enter sensory memory via the eyes and ears. WM is a limited—capacity memory system that holds information in temporary storage under attentional control when conducting a cognitive task. There are two WM models: the picture model, and the verbal model.

These models are similar to the visuospatial sketchpad and phonological loop of Baddeley's WM.

WM is involved in the process of integrating visual semantics and language. In other words, the concepts generated via the processing of audiovisual elements that are newly introduced are integrated with existing concepts supporting the provisional topic and with prior knowledge from LTM. This leads to the classification of the topic as the same or newly updated. The process of multimedia comprehension can be described as three sequential phases: attention, perception, and integration.

4.3 Relevance Model

Our proposed relevance model includes the simple search model, and the complex search model (see Figure. 4−3). The simple search model handles the process of topical relevance judgment when conducting a simple search. In this case, a clear topic search is given, for which a concrete, direct solution may exist. The complex search model handles the process of dynamic relevance judgment, wherein a search topic is not clearly defined before a search. Thus, the search topic becomes clear as the user progresses through his or her search.

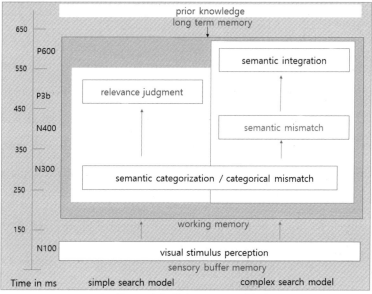

Figure 4-3. Relevance model

4.3.1 Simple Search Model

To examine the process of topical relevance judgment during a simple visual search, we assume that when a viewer rates a visual stimulus, he or she might directly relate the semantic meaning of the stimulus with a given fixed topic. For example, let T represent a given topic, and $[t_1, t_2, t_3 \cdots t_n]$ represent the semantic meanings of topic-relevant visual stimuli. The weighted values of the incoming stimuli are represented by $[s_{1w}, s_{2w}, s_{3w} \cdots s_{nw}]$, which can be measured by comparing the semantic meaning of a given stimulus with the fixed topic T, and/or by comparing that with the semantic meanings of the topic–relevant stimuli $[t_1, t_2, t_3 \cdots t_n]$. If s_{1w} is higher than a certain value, it can be classified as a topic–relevant stimulus; if it is not, then it can be classified either as a topic–partially relevant stimulus, or as a topic–irrelevant stimulus, according to its weighted value.

Logical consequences derive from this, related to ERP responses of stimuli in the simple search model. We expect that in the stimulus recognition step of the simple search model, a fronto–central N300 effect is observed because it is known to be specific to the semantic categorization of a visual stimulus, and is sensitive to categorical–level mismatches (Hamm, Johnson, & Kirk, 2002). A posterior P3b effect, which involves decision–making about whether an external stimulus matches the internal representation of a specific category, can be elicited in the relevance judgment step of the simple search model.

4.3.2 Complex Search Model

To examine the process of topical relevance judgment during a complex search, wherein the search topic becomes clear as an information seeker progresses through his or her search, we speculated that provisional or final topics are formed by comprehending incoming visual elements, inferring their meanings, and integrating their semantic meanings with the provisional topics at hand. In this, current topics seem temporary and/or local, as compared with the final global topic of a search.

This topic formation process can be expressed as follows: suppose there are a number of visual stimuli that are input in the order $[S_1, S_2, \cdots S_n]$, then the meanings of the visual stimuli, $[s_1, s_2, \cdots s_n]$ and several local and/or provisional topics, $[t_1, t_2, \cdots t_n]$ are stored in the working memory. In the real–time processing of an incoming visual stimulus during the search, several local and/or provisional topics, $[t_1, t_2, \cdots t_n]$ collaborate and/or compete with one another, in order to arrive at the final, global topic of the search, T, at the end of the search session. To this effect, we can imagine a complex dimensional matrix of relations amongst memory traces of meanings, or internal representations of visual stimuli in a participant's brain, $[s_1, s_2, \cdots s_n]$, several local and/or provisional topics, $[t_1, t_2, \cdots t_n]$, and possibly a final and global topic, T.

In terms of ERP responses to an incoming visual stimulus in the complex search model, several logical consequences become apparent. It is logical to predict that a participant will perceive a topic–irrelevant stimulus as unexpected. That is, if the participant perceives a topic–irrelevant stimulus, then the participant will tend to have a lower–negative amplitude to the topic–irrelevant stimulus than to the topic—relevant stimulus, showing the fronto—central N300 effect that is related to semantic categorization, followed by a fronto—central N400 effect. The fronto—central N400 is related to semantic mismatches that occur when a stimulus does not correspond to a prior topical expectation.

If the participant perceives a topic—relevant stimulus, then he or she tends to show the fronto—central N300/N400 effects with small negative amplitudes to the stimulus, followed by the fronto—central P600. The fronto—central P600 occurs when an incoming stimulus corresponds to prior topical expectation and its meaning needs to be integrated with prior topics.

References

Acqualagna, L., & Blankertz, B. (2013). Gaze—independent BCI—spelling using rapid serial visual presentation (RSVP). Clinical Neurophysiology, 124 (5), 901~908.

Allegretti, M., Moshfeghi, Y., Hadjigeorgieva, M., Pollick, F. E., Jose, J. M., & Pasi, G. (2015). When relevance judgement is happening?: An EEG—based Study. Proceedings of the 38th International ACM SIGIR Conference on Research and Development in Information Retrieval (pp. 719~722). New York: ACM.

Alperin, B. R., Tusch, E. S., Mott, K. K., Holcomb, P. J., & Daffner, K. R.

(2015). Investigating age-related changes in anterior and posterior neural activity throughout the information processing stream. Brain and Cognition, 99, 118~127.

Andreassi, J. L. (2006). Psychophysiology: Human behavior and physiological response. Psychology Press.

Baddeley, A. (1997). Human memory: Theory and practice (revised edition). Psychology Press.

Baddeley, A. (2006). Working memory: An overview. In S. J. Pickering (Ed.), Working memory and education (pp. 3~31). Burlington, MA: Elsevier.

Baddeley, A. (2007). Working memory, thought, and action. Oxford, UK: Oxford University Press.

Baddeley, A. (2012). Working memory: Theories, models, and controversies. Annual Review of Psychology, 63, 1~29.

Baddeley, A. (2010). Working memory. Current Biology, 20(4), R136~R140.

Baddeley, A., Eysenck, M. W., & Anderson, M. C. (2009). Memory. New York: Psychology Press.

Baddeley, A., & Hitch, G. (1974a) Working memory. Psychology of Learning and Motivation, 8, 47~89.

Baddeley, A, & Hitch, G. (1974b). Working memory. In G. Bower (Ed.), Recent advances in learning and memory (Vol. 8, pp. 47~90). New York: Academic Press.

Banich, M. T., & Compton, R. J. (2011). Cognitive neuroscience, 3rd Edition. Wadsworth.

Behneman, A., Kintz, N., Johnson, R., Berka, C., Hale, K., Fuchs, S., ... & Baskin, A. (2009, July). Enhancing text-based analysis using neurophysiological measures. In International Conference on Foundations of Augmented Cognition (pp. 449~458). Springer, Berlin, Heidelberg.

Bledowski, C., Kadosh, K. C., Wibral, M., Rahm, B., Bittner, R. A., Hoechstetter, K., ... & Linden, D. E. (2006). Mental chronometry of working memory retrieval: A combined functional magnetic resonance imaging and event-related potentials approach. Journal of Neuroscience, 26(3), 821~829.

Bledowski, C., Kaiser, J., & Rahm, B. (2010). Basic operations in working memory: Contributions from functional imaging studies. Behavioural Brain Research, 214(2), 172~179.

Bledowski, C., Prvulovic, D., Hoechstetter, K., Scherg, M., Wibral, M., Goebel, R., et al. (2004). Localizing P300 generators in visual target and distractor processing: A combined event-related potential and functional magnetic resonance imaging study. Journal of Neuroscience, 24(42), 9353~9360.

Brouwer, A. M., Hogervorst, M. A., Oudejans, B., Ries, A. J., & Touryan, J. (2017). EEG and eye tracking signatures of target encoding during structured visual search. Frontiers in Human Neuroscience, 11(264), 1~11.

Browne, P., & Smeaton, A. F. (2005, September). Video retrieval using dialogue, keyframe similarity and video objects. In Image Processing, 2005. ICIP 2005. IEEE International Conference on (Vol. 3, pp. III-1208). IEEE.

Burmester, J., Spalek, K., & Wartenburger, I. (2014). Context updating during sentence comprehension: The effect of aboutness topic. Brain and Language, 137, 62~76.

Christophel, T. B., Hebart, M. N., & Haynes, J. D. (2012). Decoding the contents of visual short-term memory from human visual and parietal cortex. Journal of Neuroscience, 32(38), 12983~12989.

Galletti, C., Kutz, D. F., Gamberini, M., Breveglieri, R., & Fattori, P. (2003). Role of the medial parieto-occipital cortex in the control of reaching and grasping movements. Experimental Brain Research, 153(2), 158~170.

Chen, Z., & Xu, Y. (2005). User-oriented relevance judgment: A conceptual model. Proceedings of the 38th Annual Hawaii International Conference on System Sciences (pp. 101b~101b).

Collette, F., Hogge, M., Salmon, E., & Van der Linden, M. (2006). Exploration of the neural substrates of executive functioning by functional neuroimaging. Neuroscience, 139(1), 209~221.

Courtney, S. M. (2004). Attention and cognitive control as emergent properties of information representation in working memory. Cognitive, Affective, & Behavioral Neuroscience, 4(4), 501~516.

Davidson, R. J., & Irwin, W. (1999). The functional neuroanatomy of emotion and affective style. Trends in Cognitive Science, 3, 11~21.

DeFrance, J. F. (1997). Age-related changes in cognitive ERPs of attenuation. Brain Topography, 9(4), 283~293.

Donchin, E., & Coles, M. (1988). Is the P300 component a manifestation of context updating? Behavioral and Brain Science, 11(3), 357~374.

Dudai, Y., (2008). Enslaving central executives: Toward a brain theory of cinema. Projections, 2(2), 21~42.

Eugster, M. J., Ruotsalo, T., Spapé, M. M., Kosunen, I., Barral, O., & Ravaja, N., et al. (2014). Predicting term–relevance from brain signals. In SIGIR 14 conference committee (Eds.), Proceedings of the 37th International ACM SIGIR Conference on Research & Development in Information Retrieval (pp. 425~434). New York: ACM Press.

Eugster, M. J., Ruotsalo, T., Spapé, M. M., Barral, O., Ravaja, N., Jacucci, G., & Kaski, S. (2016). Natural brain–information interfaces: Recomm–ending information by relevance inferred from human brain signals. Scientific Reports, 6(38580), 1~10.

Federmeier, K. D., & Laszlo, S. (2009). Time for meaning: Electrophysiology provides insights into the dynamics of representation and processing in semantic memory. In B. H. Ross (Ed.), Psychology of Learning and Motivation, 51, 1~44. Burlington: Academic Press.

Fernandes, L. S., Ferreira, D. S., Almeida, P. R., & Dias, N. S. (2015). Aging and attentional set shifting on WCST: An event–related EEG study. In Neural Engineering (NER), 2015 7th International IEEE/EMBS Conference on (pp. 1088~1091). IEEE.

Fox, S., Karnawat, K., Mydland, M., Dumais, S., & White, T. (2005). Evaluating implicit measures to improve web search. ACM Transactions on Information Systems (TOIS), 23(2), 147~168.

Funahashi, S. (2006). Prefrontal cortex and working memory processes. Neuroscience, 139(1), 251~261.

Goldenholz, D. M., Ahlfors, S. P., Hämäläinen, M. S., Sharon, D., Ishitobi,

M., & Vaina, L. M., et al. (2009). Mapping the signal-to-noise ratios of cortical sources in magnetoencephalography and electroencephalography. Human Brain Mapping, 30(4), 1077~1086.

Gwizdka, J., Mostafa, J., Moshfeghi, Y., Bergman, O., & Pollick, F. E. (2013). Applications of neuroimaging in information science: Challenges and opportunities. Proceedings of the Association for Information Science and Technology, 50(1), 1~4.

Hamann, S. B. & Squire L. R. (1997). Intact perceptual memory in the absence of conscious memory. Behavioral Neuroscience, 111, 850~854.

Hamm, J. P., Johnson, B. W., & Kirk, I. J. (2002). Comparison of the N300 and N400 ERPs to picture stimuli in congruent and incongruent contexts. Clinical Neurophysiology, 113(8), 1339~1350.

Harter, S. P. (1992). Psychological relevance and information science. Journal of the American Society for Information Science and Technology, 43(9), 602~615.

Hendahewa, C., & Shah, C. (2015). Implicit search feature based approach to assist users in exploratory search tasks. Information Processing and Management, 51(5), 643~661.

Hendahewa, C., & Shah, C. (2017). Evaluating user search trails in exploratory search tasks. Information Processing and Management, 53(4), 905~922.

Hillyard, S., Hink, R., Schwent, V., & Picton, T. (1973). Electrical signs of selective attention in the human brain. Science, 182(4108), 177~180.

Hillyard, S., & Woods, D. (1979). Electrophysiological analysis of human brain function. In M. S. Gazzaniga (Ed.), Handbook of behavioral neurobiology: Vol 2. Neuropsychology (pp. 345~378). New York: Plenum Press.

Huang, X., & Soergel, D. (2006). An evidence perspective on topical relevance types and its implications for exploratory and task-based retrieval. Information Research, 12(1), 12-1.

Ingwersen, P. (1996). Cognitive perspectives of information retrieval interaction. Journal of Documentation, 52(1), 3~50.

Jung, H. et al. (2012). Reduced source activity of event-related potentials for affective facial pictures in schizophrenia patients. Schizophrenia Research, 136(1), 150~159.

Komaki, D., Hara, T., & Nishio, S. (2012). How does mobile context affect people's web search behavior?: A diary study of mobile information needs and search behaviors. In Advanced Information Networking and Applications (AINA), 2012 IEEE 26th International Conference on (pp. 245~252). IEEE.

Kutas, M., & Federmeier, K.D. (2011). Thirty years and counting: Finding meaning in the N400 component of the event-related brain potential (ERP). Annual Review of Psychology, 62, 621~647.

Kapoor, A., Pradeep Shenoy, P., & Tan, D. (2008). Combining brain computer interfaces with vision for object categorization. IEEE Computer Society Conference on Computer Vision and Pattern Recognition (CVPR), pp. 1~8.

Kauppi, J. P., Kandemir, M., Saarinen, V. M., Hirvenkari, L., Parkkonen, L., Klami, A., ... & Kaski, S. (2015). Towards brain-activity-controlled information retrieval: Decoding image relevance from MEG signals.

NeuroImage, 112, 288~298.

Kim, H., & Kim, Y. (2010). Toward a conceptual framework of key-frame extraction and storyboard display for video summarization. Journal of the American Society for Information Science and Technology, 61(5), 927~939.

Kok, A. (2001). On the utility of P3 amplitude as a measure of processing capacity. Psychophysiology, 38(3), 557~577.

Kolb, B., & Whishaw, I. Q. (2009). Fundamentals of human neuropsychology (6th ed.). New York: Worth.

Kolb, B., & Whishaw, I. Q. (2011). An introduction to brain and behavior (3rd ed.). New York: Worth.

Kutas, M., & Hillyard, S. A. (1980). Reading senseless sentences: Brain potentials reflect semantic incongruity. Science, 207(4427), 203~208.

Lauro, L. J. R., Reis, J., Cohen, L. G., Cecchetto, C., & Papagno, C. (2010). A case for the involvement of phonological loop in sentence comprehension. Neuropsychologia, 48(14), 4003~4011.

Logie, R. H. (2011). The visual and the spatial of a multicomponent working memory. In A. Vandierendonck & A. Szmalec (Eds.), Spatial working memory (pp. 19~45). New York: Psychology Press.

Luck, S. J. (2004). Ten simple rules for designing and interpreting ERP experiments. In Handy, T.C. (Ed.), Event-related potentials: A methods handbook (pp. 17~32). Cambridge, MA: MIT Press.

Marchionini, G. (2006). Exploratory search: From finding to understanding. Communications of the ACM, 49(4), 41~46.

Martí, M., Hinojosa, J. A., Casado, P., Muñoz, F., & Fernández-Frí, C.

(2004). Electrophysiological evidence of an early effect of sentence context in reading. Biological Psychology, 65(3), 265~280.

Mayer, R. E. (2005). Cognitive theory of multimedia learning. In R.E. Mayer (Ed.), The Cambridge handbook of multimedia learning (pp. 134~146). New York: Cambridge University Press.

Moscovitch, M. (1992). Memory and working-with-memory: A component process model based on modules and central systems. Journal of Cognitive Neuroscience, 4, 257~267.

Moshfeghi Y., Pinto, L.R., Pollick, F.R., & Jose, J.M. (2013). Understanding relevance: An fMRI study. In P. Serdyukov et al., eds. Advances in Information Retrieval. Springer Berlin Heidelberg.

Mudrik, L., Shalgi, S., Lamy, D., & Deouell, L. Y. (2014). Synchronous contextual irregularities affect early scene processing: Replication and extension. Neuropsychologia, 56, 447~458.

Naatanen, R, Simpson, M., & Loveless, N. (1982). Stimulus deviance and evoked potentials. Biological Psycology, 14, 53~98.

Naghavi, H., & Nyberg, L. (2005). Common fronto-parietal activity in attention, memory, and consciousness: Shared demands on integration? Consciousness and Cognition, 14(2), 390~425.

Nakano, H., Rosario, M. A. M., Oshima-Takane, Y., Pierce, L., & Tate, S. G. (2014). Electrophysiological response to omitted stimulus in sentence processing. NeuroReport, 25(14), 1169~1174.

Picton, T. (1988). The endogenous evoked potentials. In: E. Basar (Ed.) Dynamics of sensory and cognitive processing by the brain (pp. 258~265). Springer-Verlag, Berlin.

Picton, T., Campbell, K., Barlbeau−Beaun, J., & Proulx, G. (1978). The neurophysiology of human attention: A tutorial review. In J. Requin (Ed). Attention and performance: VII. (pp. 429~467). New York: Wiley.

Postle, B. R., Stern, C. E., Rosen, B. R., & Corkin, S. (2000). An fMRI investigation of cortical contributions to spatial and nonspatial visual working memory. NeuroImage, 11(5), 409~423.

Reuter−Lorenz, P. A., & Jonides, J. (2007). The executive is central to working memory: Insights from age, performance, and task variations. In A. R. A. Conway et al. (Eds.), Variations in working memory (pp. 250~271). New York: Oxford University Press.

Repovš, G., & Baddeley, A. (2006). The multi−component model of working memory: Explorations in experimental cognitive psychology. Neuro−science, 139(1), 5~21.

Rieh, S. Y., Collins−Thompson, K., Hansen, P., & Lee, H. J. (2016). Towards searching as a learning process: A review of current perspectives and future directions. Journal of Information Science, 42(1), 19~34.

Rudner, M., Fransson, P., Ingvar, M., nyberg, l., & Rönnberg, J. (2007). Neural representation of binding lexical signs and words in the episodic buffer of working memory. Neuropsychologia, 45, 2258~2276.

Ruotsalo, T., Jacucci, G., Myllymäki, P., & Kaski, S. (2015). Interactive intent modeling: Information discovery beyond search. Communica−tions of the ACM, 58(1), 86~92.

Saracevic, T. (2007). Relevance: A review of the literature and a framework for thinking on the notion in information science. Part II: Nature and manifestations of relevance. Journal of the Association for Information Science and Technology, 58(13), 1915~1933.

Seoane, L. F., Gabler, S., & Blankertz, B. (2015). Images from the mind: BCI image evolution based on rapid serial visual presentation of polygon primitives. Brain—Computer Interfaces, 2(1), 40~56.

Singer, G., Danilov, D., & Norbisrath, U. (2012). Complex search: Aggregation, discovery, and synthesis. Proceedings of the Estonian Academy of Sciences, 61, 89~106.

Singer, G., Norbisrath, U., & Lewandowski, D. (2013). Ordinary search engine users carrying out complex search tasks. Journal of Information Science, 39(3), 346~358.

Sitnikova, T., Holcomb, P. J., Kiyonaga, K. A., & Kuperberg, G. R. (2008). Two neurocognitive mechanisms of semantic integration during the comprehension of visual real—world events. Journal of Cognitive Neuroscience, 20(11), 2037~2057.

Smith, M. E. (1993) Neurophysiological manifestations of recollective experience during recognition memory judgments. Journal of Cognitive Neuroscience, 5, 1~13.

Smith, E. E., & Jonides, J. (1999). Storage and executive processes in the frontal lobes. Science, 283(5408), 1657~1661.

Soltani, M., & Knight, R. T. (2000). Neural origins of the P300. Critical ReviewsTM in Neurobiology, 14(3–4), 199~224.

Thatcher, R. W., Krause, P. J., & Hrybyk, M. (1986). Cortico—cortical associations and EEG coherence: A two—compartmental model. Elect—

roencephalography and Clinical Neurophysiology, 64(2), 123~143.

Thornhill, D. E., & Van Petten, C. (2012). Lexical versus conceptual anticipation during sentence processing: Frontal positivity and N400 ERP components. International Journal of Psychophysiology, 83(3), 382~392.

Tulving, E. et al. (1994). Hemispheric encoding / retrieval asymmetry in episodic memory: Position emission tomography findings. Proceedings of the National Academy of Sciences of the United States of America, 91(6), 2016~2020.

van Berkum, J. J., Brown, C. M., Zwitserlood, P., Kooijman, V., & Hagoort, P. (2005). Anticipating upcoming words in discourse: Evidence from ERPs and reading times. Journal of Experimental Psychology: Learning, Memory, and Cognition, 31(3), 443.

van Berkum, J. J., Hagoort, P., & Brown, C. M. (1999). Semantic integration in sentences and discourse: Evidence from the N400. Journal of Cognitive Neuroscience, 11(6), 657~671.

Ventre-Dominey, J., Bailly, A., Lavenne, F., Lebars, D., Mollion, H., Costes, N., & Dominey, P. F. (2005). Double dissociation in neural correlates of visual working memory: A PET study. Cognitive Brain Research, 25(3), 747~759.

Verleger, R. (1988). Event-related potentials and cognition: A critique of the context updating hypothesis and an alternative interpretation of P3. Behavioral and Brain Sciences, 11(3), 343~356.

Verleger, R. (2008). P3b: Towards some decision about memory. Clinical Neurophysiology, 119(4), 968~970.

Verleger, R., Jaśkowski, P., & Wascher, E. (2005). Evidence for an integrative role of P3b in linking reaction to perception. Journal of Psychophysiology, 19(3), 165~181.

Võ, M. L. H., & Wolfe, J. M. (2013). Differential ERP signatures elicited by semantic and syntactic processing in scenes. Psychological Science, 24(9), 1816.

Wang, L., & Schumacher, P. B. (2013). New is not always costly: Evidence from online processing of topic and contrast in Japanese. Frontiers in Psychology, 4, 363.

Wang, S., Zhu, Y., Wu, G., & Ji, Q. (2014). Hybrid video emotional tagging using users' EEG and video content. Multimedia Tools and Applications, 72(2), 1257~1283.

Wang, Y. (2009). A cognitive informatics theory for visual information processing, Proc. 7th IEEE International Conference on Cognitive Informatics (ICCI'08), Stanford University, CA.

West, W. C., & Holcomb, P. J. (2002). Event-related potentials during discourse-level semantic integration of complex pictures. Cognitive Brain Research, 13(3), 363~375.

Xu, X., & Zhou, X. (2016). Topic shift impairs pronoun resolution during sentence comprehension: Evidence from event-related potentials. Psychophysiology, 53(2), 129~142.

Zhu, X., Goldberg, A. B., Eldawy, M., Dyer, C. R., & Strock, B. (2007). A text-to-picture synthesis system for augmenting communication. Association for the Advancement of Artificial Intelligence, 7, 1590~1595.

Chapter 5

Metadata Framework for Efficient
Browsing and Searching of Web Videos

5. Metadata Framework for Efficient Browsing and Searching of Web Videos

We introduce three multimedia metadata formats, PBCore, MPEG–7, and TVA in this chapter, and also detail the metadata framework for efficient browsing and searching of Web videos, which was proposed by the author.

5.1 Introduction

The advances in information technology together with the rapid evolution of multimedia data are resulted in the significant growth of Web videos. Owing to such rapid growth of Web videos over the Internet, it is becoming extremely important to perform accurate and efficient content–based searches and to provide nonlinear access to any segment of them without having to view them in their entirety. Furthermore, there is a strong demand for short pieces of AV content in the archive of media professionals or video users (Huurnink et al., 2010; Yang & Marchionini, 2004).

When users search for videos, some of them only want a small part of a video, but usually they need to go through the entire video to find it. To solve this problem, videos need to be indexed at the scene or sequence level through automatic segmentation. Hence, we need to employ a structural and semantic metadata framework; structural metadata is used to describe the structure of AV content in terms of video segments, frames, still and moving regions, and audio segments. The semantic metadata to represent the objects, events, and notions from the real world are captured by the AV content. MPEG−7 supports multimedia structural and semantic descriptions. However, it is known that MPEG−7 is not currently suitable for describing multimedia content on the Web (Arndt et al., 2007) and furthermore, it is too complicated to work with (List & Fisher, 2004).

In this study, we proposed a structural and semantic metadata framework, which was designed by taking a Web environment into consideration. Specifically, in order to investigate how to structure Web videos and which structural and semantic metadata elements are useful, we first reviewed social metadata and three international multimedia metadata formats, namely PBCore 2.0, MPEG−7, and TVA.

Then, to identify the events in AV content, we adapted Chatman's narrative theory (Chatman, 1975) in which a narrative is any report of connected events, presented in a sequence of written words or videos (Teeter & Sandberg, 2016; Chatman, 1975).

Comprehending an event depends on identifying the nature of its key action and the roles played by the people and objects in the action (Nowak, Plotkin, & Jansen, 2000; Klix, 2001). Thus, we assume that an event is an instantaneous or temporally extended action or happening that may involve a single agent object or object, an interaction between two or more agent objects, or all of the agent objects (Shotton et al., 2002).

Therefore, for the elements of event specific metadata, we used object, agent object, action, place, time, and theme extracted from the analysis of a discourse such as dialogues, sound, and music. Our proposed metadata framework, constructed through the abovemen-tioned steps, can be utilized for applications such as content-based searches, fast browsing, and dynamic video summarization.

5.2 Multimedia Metadata: A Comparison of PB-Core, TVA, and MPEG-7

Below, we describe three international multimedia metadata formats, namely PBCore 2.0 (http://pbcore.org/introducing-pbcore-2-0/), MPEG-7 (ISO/IEC, 2002~2004) (http://mpeg.chiariglione.

org/standards/mpeg-7), and TVA (http://www.tv-anytime.org/) (Evain & Martínez, 2007). We selected three multimedia metadata formats because they are international standards and used by many digital libraries. PBCore 2.0 was created by the public broadcasting community in the United States of America for use by public broadcasters. PBCore 2.0 was built on the foundation of the Dublin Core (ISO 15836) and is made up of 4 content classes, 15 containers, and 82 elements. The four content classes consist of intellectual content (descriptive metadata), intellectual property (creation and usage information), instantiation (technical metadata), and extension.

TVA includes content description metadata, instance description metadata, segmentation metadata, and consumer metadata. The content description metadata describes general information about a piece of content (e.g., title), whereas the instance description metadata refers to a particular instance of a piece of content (e.g., video format) (Lee et al., 2005). Segmentation metadata is used to define, access, and manipulate temporal intervals (e.g., segments) within an AV stream and the consumer metadata includes usage history data and user preferences for a personalized content service.

MPEG-7 consists of 12 parts including multimedia description schemes (MDSs), AV, and systems. Among them, MDSs, audio, and visual are the most important parts of it. MDSs, which are metadata structures for describing and annotating AV content, include basic elements (schema tools, basic datatypes, link and media localization, and basic tools), content description, content management, content organization, navigation and access, and user interaction (Salembier & Smith, 2001).

Content description represents the structure and semantics of AV content. That is, the content description tools represent perceivable information, comprising structural aspects (structural description tools) and conceptual aspects (semantic description tools). The structural description tools allow the description of the content in terms of spatio−temporal segments organized in a hierarchical structure. The semantic description tools allow the description of the content from the viewpoint of real−world semantics and conceptual notions: objects, events, abstract concepts, and relationships. The semantic and structural description tools can be further related by a set of links, which allow AV content to be described on the basis of both content structure and semantics together.

Further, visual covers basic visual features such as color, texture, shape, motion, localization, and face recognition, whereas audio provides structures for describing audio content.

As described above, three metadata formats (PBCore 2.0, MPEG−7, and TVA) have in common the fact that they have many overlapping metadata elements, and that they do not consider social relationships among users and social metadata.

Despite many similarities among them, there are some differences in the following aspects. First, MPEG–7 enables the specification of low–level AV features such as color and motion, in addition to the specification of high–level AV features such as segments, objects, and events. Those features are useful for applications such as content–based querying and video summarization.

Second, TVA focuses more on describing consumer profiles including search preferences to facilitate automatic filtering and the acquisition of content by agents on behalf of the consumer, compared to the other two metadata formats. Third, PBCore 2.0 provides a simple way to describe AV content.

We compared these three metadata formats in terms of creation and production information, media information, structural and semantic information, and usage information (see Table 5–1). The creation and production information are related to data about the creation and production of AV content, media information is related to the media–specific characteristics of the content, and the usage information to the usage of the content, such as copyright and usage history.

Further, the structural information describes the AV content from the viewpoint of its structure, whereas the semantic information represents the content from the viewpoint of real–world semantics and conceptual notions

Table 5-1. A comparison of PB-Core (2.0), TVA (1.3), and MPEG-7 (10)

Creation and Production Information		
PB-Core(2.0) (2011)	TVA(1.3) (2003)	MPEG-7(10) (2001)
pbcorePublisher	programDescription	abstract
(publisher, publisherRole)	keyword	language
pbcoreSuject	genre	captionLanguage
pbcoreDescription	language	signLanguage
pbcoreAnnotation	captionLanguage	relatedMaterial DS
pbcoreGenre	signLanguage	(publicationType,
pbcorecoverage	relatedMaterial	materialType,
(coverage, coverageType)		mediaLocator,
pbcoreRelation		mediaInformation,
(pbcoreRelationType,		creationInformation,
pbcoreRelationIdentifier)		usageInformation)
Media Information		
PB-Core	TVA	MPEG-7
pbcoreInstantiation	fileFormt	fileFormt
(instantiationIdentifier,	fileSize	fileSize
instantiationLocation,	system	bitRate
instantiationDuration,	bitRate	visualCoding
instantiationMediaType,	audioAttributes	audioCoding
instantiationColors, etc.)	videoAttributes	visual
instantiationEssenceTrack		(color, texture, shape,
(essenceTrackType,		motion, localization, and
essenceTrackAspectRatio,		face recognition)
essenceTrackEncoding,		audio
essenceTrackFrameRate,		(silence, spoken content,
etc.)		timbre, sound effects,
		melody contour, etc.)
Structural and Semantic Information		
PB-Core	TVA	MPEG-7
pbcorePart	segment (title, synopsis,	structure DS
(pbcoreIdentifier,	keyword, relatedMaterial,	segment DS (creation

pbcoreTitle, pbcoreDescription, pbcoreSubject, etc.)	programRef, description, segmentLocator, keyFrameLocator) segmentGroup (programRef, groupType, description, groupInterval, segments, groups, keyFrameLocator)	information, usage information, media information, textual annotation) segmentRelation DS semantic DS (object DS, agentObject DS, event DS, concept DS, semanticState DS, semanticPlace DS, semanticTime DS)
Usage Information		
PB-Core	TVA	MPEG-7
pbcoreAudienceLevel pbcoreAudienceRating pbcoreRightsSummary (rightsSummary, rightsLink, rightsEmbedded)	releaseInformation (releaseDate, releaseLocation) parentalGuidance userDescription (userPreference, usageHistory) awardsList copyrightNotice	release (country, date) target (market, age) usageInformation DS (rights, availability, usageRecord, etc.)

5.3 Related Studies

Cunningham and Nichols (2008) suggested that the video queries submitted by many participants were driven by their mood or emotional state, and YouTube was the primary site they consulted. Huurnink et al. (2010) mentioned that media professionals have a strong demand for short pieces of AV material in the archive, and their queries predominantly consist of broadcast titles and of proper names.

Shotton et al. (2002) proposed a metadata classification schema for the characterization of items and events in biological cell videos that permit a subsequent query by content. Following MPEG–7 nomenclature, they first defined metadata intrinsic to the information content of the video as either structural metadata or semantic metadata. Then, they showed how the semantic metadata types should be organized within a database.

Agnew, Kniesner, and Weber (2007) described the implement–ation of MPEG–7 within the Moving Image Collections, which is a union catalog of the world's moving images. They also discussed issues with MPEG–7 as a descriptive metadata schema, as well as mapping and implementation issues identified in their study.

Benitez, Zhong, and Chang (2007) proposed two research prototype systems that demonstrate the generation and consumption of MPEG–7 structure and semantic descriptions in retrieval applications. The first prototype system allows to segment and model objects as a set of regions with corresponding visual features and spatiotemporal relationships. The region–based model provides an effective base for the retrieval of objects based on similarity. Whereas the second system enables the representation of semantic and perceptual facts about the world using multimedia.

Lunn (2009) investigated three aspects of scholars' and students' information seeking behavior in a television broadcast context, and the associated implications for the design and construction of metadata elements in surrogate records in future broadcast retrieval systems. The three aspects of information seeking behavior in focus are information requirement characteristics, preferred search entries, and the application of relevance criteria.

His research results demonstrated that 24 access points (metadata elements) are appropriate in relation to a future broadcast retrieval system. These are title, participant, author, date of production, channel, summary, spoken words, clips, and images. He mentioned that in the questionnaire, respondents expressed the need for a further specification of roles for the author access point. Roles in such a specification are director, actor, cameraman, commentator, and anchorperson.

Makkonen et al. (2010) proposed two algorithms that are used to find events by using video metadata. The first algorithm deals with missing data compensation that harvests missing data values from textual descriptions in video metadata. The second algorithm is a layered clustering method that divides videos into clusters, each of which is considered as an event. Their test results of the two methods showed that the missing data compensation yielded better results in terms of accuracy than using raw data, and that the clustering method can produce good quality clusters of events.

Algur, Bhat, and Jain (2014) constructed high–level descriptive metadata for all categories of Web videos such as those on YouTube. They suggested that by using high–level descriptive metadata information, a user is facilitated in locating a specific video and is further able to rapidly comprehend the main concept of a video without the need to watch it in its entirety. They proposed general descriptive metadata, object–specific metadata, and event–specific metadata. General descriptive metadata have 13 elements including title, director, actors, theme, language, storage format, number of emotional scenes, and number of songs.

Conversely, the event–specific metadata consisted of seven elements including category, time, and starting frame, whereas the object–specific metadata have seven elements such as object ID and motion of the object.

Cho (2014) analyzed the structure of European Broadcasting Union Core, PB Core, and Korean Broadcasting System (KBS) metadata, and found no big differences among the three formats. He suggested that for efficient browsing and reusing of AV content, the AV content needs to be segmented based on shots or scenes rather than sequences, namely a series of connected scenes.

Taking into account the findings and issues from the previous studies, we can conclude that it may be beneficial to allow fine-grained access to AV material through automatic segmentation or content-based analysis. For this end, Shotton et al. (2002) proposed a good semantic metadata schema, but it was designed for biological cell videos and thus domain-specific. Algur at al. (2014) described general descriptive metadata, object-specific metadata, and event-specific metadata for general Web videos. However, they did not provide a framework for describing the relationships among these three metadata. Therefore, we believe that a structural and semantic metadata framework design is required for general Web videos to be related in a more elaborate way.

5.4 Structural and Semantic Metadata Framework

5.4.1 Theoretical Model

To construct an efficient structural and semantic metadata framework for video data, we need to examine how the narrative of a video is constructed and how its topic is determined by viewers. We reviewed Chatman's narrative theory (Chatman, 1975) in order to precisely examine the concept of a narrative (see Chapter 2).

5.4.2 Metadata Framework

Below, we describe our proposed structural and semantic metadata framework that was constructed on the basis of Chatman's narrative theory, social data, and three multimedia metadata formats. Humans often claim to understand events when they manage to formulate a coherent story or narrative explaining how they believe the event was generated; thus, narratives lie at the foundation of our cognitive procedures.

On the basis of Chatman's narrative theory, our metadata framework includes event and eventGroup information in order to divide a video into clusters, each of which is considered as an event. We also include both segment and video (program) information that are closely related to the event information.

We utilized social data, such as social metadata, social content, and social relation information being obtained from social network services, such as Facebook and YouTube. Social data are useful in that they enable to automatically construct metadata and to use social relation information. However, social data have some disadvantages such as spam tags; thus, it is necessary to filter them before their use.

1) Segment Information

A program can be segmented automatically based on a shot, a

scene, or a sequence. Any segment may be described by metadata elements that are used to represent a whole video, because segments are parts of a video. Moreover, a specific feature for a segment is also allowed. The specific feature is segmentLocator, which describes the location of a segment within a program in terms of start time, start frame number, and end frame number.

2) Event and eventGroup Information

Our metadata framework has event and eventGroup information in order to divide a program into events. For the elements of an event specific metadata (see Figure 5−1 and Table 5−2), there are 12 elements including object, agentObject, action, place, time, and theme extracted from the analysis of a discourse such as dialogues, sound, and music. An event can be linked to a segment or segments through its segmentID (s).

Then, we created an eventGroup that denotes a collection of events that are grouped together, for a particular purpose or due to a shared property. An eventGroup contains events, or other event Groups. For example, if there is a birthday party, that is an important event in the overall story of a program. Then the agent Object, action, place, and time elements can be attached to the birthday party event. Depending on the type of event, we can group events extracted from either a program or several programs through the eventGroup information.

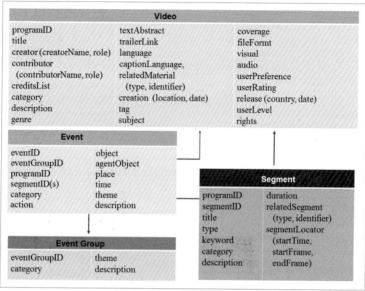

Figure 5-1. Structural and semantic metadata framework

Table 5-2. Example of the proposed metadata framework construction
(key metadata elements are described)

Video		
Element	Description	Example
programID	An ID of a given AV material	PID: 012
title	A name or label relevant to a material	Land
creator (creatorName, role)	creatorName and role sub–elements identify the primary person (s), or organization(s) responsible for creating a material and the role played by the creator (s)	creator (H. Kim, director)

contributor (contributorName, role)	contributorName and role sub−elements identify person (s), or organization(s) that made substantial creative contributions to a material and the role played by the contributor (s)	contributor (H. Park, narrator)
creditsList	The list of credits (e.g. actors) for a material	Actors: S. kim …
category	Category of a material	Shows
trailerLink	A URI pointing to the dynamic video skim for a material	http://www …
relatedMaterial (type, identifier)	Type sub−element describes the relationship between the AV material being described and any other AV material, and identifier sub−element indicates an ID of a related AV material or an URL that points to it.	relatedMaterial (part−whole relations, PID1234)
creation (location, date)	Location and date sub−elements identify where and when a material was produced (created)	creation (Seoul, 2015)
visual/audio	Information for basic visual features (e.g., color, or object recognition) and audio content (e.g., melodic contour)	Visual and audio information
userPreference	User preference information, such as favorite actors or TV shows	Information about favorite video genre
userRating	A user's rating for an AV material	Number of likes and dislikes
release (country, date)	Country and date sub−elements identify where and when a material was released	release (Korea, 2015)
userLevel	A type of audience for whom a material is primarily designed or educationally useful	For educational use (K−2)
Segment		
Element	Description	Example
programID	An ID of the AV material that includes a given segment	PID: 2354

segmentID	An ID of a segment	SID: 230
title	A name or label relevant to a segment	Birthday party
type	Type of video segmentation	Shots or scenes
relatedSegment (type, identifier)	Type sub-element describes the relationship between the segment being described and any other segment, and identifier indicates an ID of a related segment or an URL that points to it.	relatedSegment (part-whole relations, SID123)
segmentLocator (startTime, startFrame, endFrame)	Location of the segment within a program in terms of start time, start frame number, and end frame number	segmentLocator (00:05:54, 0, 183)

Event		
Element	Description	Example
eventID	An ID of the event described	EID: 34
programID	An ID of the AV material that includes the event	PID: 12
segmentID(s)	ID (s) of the segment (s) that includes the event	SID: 12
category	Category for the event	Entertainment
action	Action described in the event	Dancing
object	Object appeared in the event	Flowers
agentObject	Agent object appeared in the event	Six people
place	Place related to the event	Park
time	Time related to the event	01:12: 2017
theme	Topic for the event	Social sports

Event Group		
Element	Description	Example
eventGroupID	An ID of grouped events	GEID: 23
category	Category for grouped events	Entertainment
theme	Topic for grouped events	Comedy

3) Video Information

A whole video has 25 metadata elements that consist of creation and production information, media information, usage information, and social information (e.g., tag). Among them, the relatedMaterial element contains two sub-elements: type and identifier. The type element is used to describe the relationship between the AV material being described and any other AV material; both AV materials can be related by part-whole relations or different versions of an original.

Furthermore, a visual element is used to describe basic visual features such as color, texture, shape, motion, localization, face recognition, and object recognition, whereas an audio element is used to describe audio content, such as spoken content, timbre, and sound effects. These visual and audio elements are employed to facilitate content-based searching (e.g., query-by-humming).

5.5 How to Obtain Metadata Information

There are several ways to obtain metadata in addition to metadata that are manually entered by metadata mangers (or librarians) (Christel, 2009; Park & Lu, 2009). As in YouTube, some descriptive metadata including titles and tags can be obtained from video uploaders or content creators. In this case, if it is not required to input metadata elements, we cannot obtain enough metadata.

We are also able to generate metadata automatically using automated content analysis tools. Some technical metadata such as duration and number of frames can be generated by employing video content analysis tools, such as MediaInfo (https://en.wikipedia. org/wiki/MediaInfo) and MooO (http://www.moo0.com/); for example, MediaInfo reveals metadata information, such as title,

Figure 5-2. Technical metadata obtained through MediaInfo

author, director, duration, codec, framerate, and subtitle language (Figure 5-2). Video trailers can be generated automatically by using Azure Media Services (https://azure. microsoft.com/), which can help create a video skim of a long video by automatically extracting key shots from the video.

For the construction of a video summary, video shot (or scene) boundary detection techniques that can be used to automatically segment a video have received attention in the field of computerized image processing, and the techniques have made a large amount of progress (Smeaton, Over, & Doherty, 2010; Chen, Delannay, & Vleeschouwer, 2011).

Furthermore, the content features of a video such as faces and objects can be also identified automatically by using image processing techniques (Togawa & Okuda, 2005; Yokoi, Nakai, & Sato, 2008). However, such content feature extraction techniques have not yet been developed to the point where they are robust or effective in dealing with large collections.

Therefore, it is necessary to review other techniques and theories utilized together with image processing techniques. One such area is theories of cognitive psychology and cognitive neuroscience techniques. Behroozi, Daliri, and Shekarchi (2016) described that visual attention-related brain activities evoked by stimulus can be regarded as the signature of detection and identification of objects.

That is, they investigated the possibility of identifying a conceptual representation based on the presentation of 12 semantic categories (e.g., animal and building) of objects using EEG signals in conjunction with a multivariate pattern recognition technique.

Lastly, to obtain metadata information, we can use social metadata (e.g., tags), social content, and social relation information. Users have been provided the capability of describing documents or videos with subject headings or "tags." This tagging reflects the users' conceptual model of information, and the tags present authentic representations of the language of authors and users (Quintarelli, 2005; Peters & Stock, 2007).

For the userPreference element that is used to describe a user's preferences for consumption of multimedia material (see Figure 5–1), we can utilize two social recommendation methods: a collaborative filtering approach and a content-based approach (Bouadjenek, Hacid, & Bouzeghoub, 2016). The collaborative filtering approach intends to recommend items to a user based on other people who are found to have similar preferences or tastes, whereas the content-based approach is based on recommending items that are similar to those in which the user has shown interest in the past.

Hence, the personalization of content access and content consumption can be done by leveraging user preference information, which will be matched with media descriptions and also processed by using other people's preference information if they have similar user preferences.

For the userRating element that is used to describe a user's rating for an AV material (see Figure 5-1), we are able to leverage the social data assigned by users, such as the number of likes and dislikes of a material and comments about a material.

5.6 Conclusion

With the explosive growth of Web videos on the Internet, it becomes challenging to efficiently browse and search hundreds or even thousands of Web videos. In this study, we proposed a structural and semantic metadata framework for Web videos after reviewing three metadata formats (PBCore 2.0, MPEG-7, and TVA), social metadata, and related studies.

Our metadata framework allows to structure a program (narrative) on the basis of events, and it consists of event information, eventGroup information, segment information, and video (program) information. Social data being obtained from social network services are utilized for some elements of the video information.

We speculate that our proposed metadata framework can be utilized to perform content-based searches, fast browsing, or relevance assessment, and to construct static video summaries (storyboards) or dynamic video skims (trailers).

References

Agnew, G., Kniesner, D., & Weber, M. B. (2007). Integrating MPEG-7 into the Moving Image Collections portal. Journal of the Association for Information Science and Technology, 58(9), 1357~1363.

Algur, S. P., Bhat, P., & Jain, S. (2014). Metadata construction model for Web videos: A domain specific approach. International Journal of Engineering and Computer Science, 3(12), 9742~9748.

Arndt, R., Troncy, R., Staab, S., Hardman, L., & Vacura, M. (2007). COMM: Designing a well-founded multimedia ontology for the web. In The semantic web (pp. 30~43). Springer, Berlin, Heidelberg.

Behroozi, M., Daliri, M. R., & Shekarchi, B. (2016). EEG phase patterns reflect the representation of semantic categories of objects. Medical & Biological Engineering & Computing, 54(1), 205~221.

Benitez, A. B., Zhong, D., & Chang, S. F. (2007). Enabling MPEG-7 structural and semantic descriptions in retrieval applications. Journal of the Association for Information Science and Technology, 58(9), 1377~1380.

Bouadjenek, M. R., Hacid, H., & Bouzeghoub, M. (2016). Social networks and information retrieval, how are they converging? A survey, a taxonomy and an analysis of social information retrieval approaches and platforms. *Information Systems, 56,* 1~18.

Chatman, S. (1975). Towards a theory of narrative. *New Literary History, 6* (2), 295~318.

Chen, F., Delannay, D., & De Vleeschouwer, C. (2011). An autonomous framework to produce and distribute personalized team–sport video summaries: A basketball case study. *IEEE Transactions on Multimedia, 13*(6), 1381~1394.

Cho, Y. (2014). The study on improvement of broadcast metadata about clip video at broadcast content managements. In *Proceedings of 2014 Korean Society of Broadcast Engineers Summer Conference*, Jeju, South Korea.

Christel, M. G. (2009). Automated metadata in multimedia information systems: Creation, refinement, use in surrogates, and evaluation. *Synthesis Lectures on Information Concepts, Retrieval, and Services, 1*(1), 1~74.

Cunningham, S. J., & Nichols, D. M. (2008, June). How people find videos. In *Proceedings of the 8th ACM/IEEE–CS Joint Conference on Digital Libraries* (pp. 201~210). ACM.

Evain, J. P., & Martínez, J. M. (2007). TV-Anytime Phase 1 and MPEG-7. *Journal of the Association for Information Science and Technology, 58*(9), 1367~1373.

International Organization for Standardization/International Electrotechnical Commission (ISO/IEC). 2002, 2003, 2004. ISO/IEC. 15938. Part 1~8. *Information technology—Multimedia content description interface* (MPEG-7).

Huurnink, B., Hollink, L., Van Den Heuvel, W., & De Rijke, M. (2010). Search behavior of media professionals at an audiovisual archive: A transaction log analysis. Journal of the Association for Information Science and Technology, 61(6), 1180~1197.

Kim, Y. (2009). A structural model of mediated visual communication in narrative movies: Focusing on Chatman and Bordwell's controversy. Korean Journal of Journalism and Communication Studies, 53(1), 209~232.

Klix, F. (2001). The evolution of cognition. Journal of Structural Learning and Intel. Systems, 14, 415~431.

Lee, H., Yang, S. J., Lee, H. K., & Hong, J. (2005). Personalized TV services and t-learning based on TV-Anytime metadata. Advances in Multimedia Information Processing-PCM 2005, 212~223.

List, T., & Fisher, R. B. (2004, August). CVML-an XML-based computer vision markup language. In Pattern Recognition, 2004. ICPR 2004. Proceedings of the 17th International Conference on (Vol. 1, pp. 789~ 792). IEEE.

Lunn, B. K. (2009). Towards the design of user based metadata for television broadcasts: Investigation of metadata preferred in searching and assessing relevancy of television broadcasts: Emphasis on scholars and students in media studies. Saarbr¨ucken, Germany: VDM.

Makkonen, J., Kerminen, R., Curcio, I. D., Mate, S., & Visa, A. (2010, October). Detecting events by clustering videos from large media databases. In Proceedings of the 2nd ACM International Workshop on Events in Multimedia (pp. 9~14). ACM.

Mehmood, I., Sajjad, M., Rho, S., & Baik, S. W. (2016). Divide-and-conquer based summarization framework for extracting affective video content. Neurocomputing, 174, 393~403.

MPEG Home Page. [online] [cited 2016. 9. 11.] <http://mpeg.chiariglione.org/>

Nowak, M. A., Plotkin, J. B., & Jansen, V. A. (2000). The evolution of syntactic communication. Nature, 404(6777), 495~498.

Park, J. R., & Lu, C. (2009). Application of semi-automatic metadata generation in libraries: Types, tools, and techniques. Library & Information Science Research, 31(4), 225~231.

PBCore Home Page. [online] [cited 2016. 9. 3.] http://pbcore.org/introducing-pbcore-2-0/

Peters, I., & Stock, W.G. (2007). Folksonomy and information retrieval. Proceedings of the ASIST Annual Meeting, 44, 1~18.

Quintarelli, E. (2005, June). Folksonomies: Power to the people. Paper presented at the ISKO Italy UniMIB meeting, Milan, Italy.

Reijnders, K. (2011). Suspense Tours: Narrative generation in the context of tourism. Thesis Master Information Studies Programme of Human Centered Multimedia, Universiteit van Amsterdam.

Salembier, P., & Smith, J. R. (2001). MPEG-7 multimedia description schemes. IEEE Transactions on Circuits and Systems for Video Technology, 11(6), 748~759.

Shotton, D. M., Rodriguez, A., Guil, N., & Trelles, O. (2002). A metadata classification schema for semantic content analysis of videos. Journal of Microscopy, 205(1), 33~42.

Smeaton, A. F., Over, P., & Doherty, A. R. (2010). Video shot boundary detection: Seven years of TRECVid activity. Computer Vision and Image Understanding, 114(4), 411~418.

Teeter, P., & Sandberg, J. (2017). Cracking the enigma of asset bubbles with narratives. Strategic Organization, 15(1), 91~99.

Togawa, H., & Okuda, M. (2005, July). Position-based keyframe selection for human motion animation. In Parallel and Distributed Systems, 2005. Proceedings. 11th International Conference on (Vol. 2, pp. 182~185). IEEE.

TV Anytime Home Page. [online] [cited 2016. 5. 16.] <http://www.tv-anytime.org/>

Wang, M., Hong, R., Li, G., Zha, Z. J., Yan, S., & Chua, T. S. (2012). Event driven web video summarization by tag localization and key-shot identification. IEEE Transactions on Multimedia, 14(4), 975~985.

Yang, M., & Marchionini, G. (2004). Exploring users' video relevance criteria—A pilot study. Proceedings of the Association for Information Science and Technology, 41(1), 229~238.

Yokoi, K., Nakai, H., & Sato, T. (2008). Toshiba at TRECVID 2008: Surveillance Event Detection Task. In TRECVID.

Chapter 6

Speech Summarization

6. Speech Summarization

In an era of big data, with huge repositories of structured, semi—structured, or unstructured data (Cumbley & Church, 2013; Stanton, 2012), we need to analyze and summarize large amounts of speech data, even in the absence of transcripts and metadata. Unstructured lecture speech videos contain enormous information in their audio components. However, multimedia, such as videos and images, does not contain much baked—in metadata (Smith, 2008). Owing to the lack of proper surrogates and metadata, unstructured lecture videos are often not used to their full potential as educational support materials.

An abstract can be defined as the form of quite short texts or multimedia (e.g., still images, moving images, and sound) either accompanying the original material or included in its surrogate. An abstract is different from an extract, in that an abstract is a piece of material created by the abstractor, whereas an extract is an abbreviated version of a material created by drawing parts from the material itself (Chowdhury, 2010). Summaries can be either generic summaries or query—relevant summaries (Gong & Liu, 2001;

Mendoza et al., 2014; Abdi et al., 2017). A generic summary provides an overall sense of the document's contents. A good generic summary should contain the main topics of the document while keeping redundancy to a minimum. On the other hand, a query-based summary is a specific kind of a document summary. Given a user query, the task is to produce from the document a summary which can provide informative information corresponding to the user's information needs.

Extractive speech summarization is a process that selects salient parts of a speech to form a summary. Speech summarization utilizes supervised and unsupervised methods (Patil & Brazdil, 2007; Liu & Hakkani−Tur, 2011). Additionally, for extractive generic summariz−ation of transcribed lecture videos, viewer−assigned keywords (tags) were used because they reflect an author's conceptual model of information (Kim & Kim, 2016; Quintarelli, 2005; Peters & Stock, 2007). We regard this tag−based summarization as social summ−arization.

We review in this chapter previous studies on three speech summarization methods, such as supervised, unsupervised, and social summarization methods. It also details four speech summarization methods: social summarization, latent semantic analysis (LSA), maximum marginal relevance (MMR), and acoustic methods.

6.1 Related Studies

We review previous studies on supervised, unsupervised, and social summarization (see Table 6−1).

6.1.1 Supervised Summarization

Supervised techniques use lexical, structural, and discourse attributes; they rely on data sets labeled by human annotators. Maskey and Hirschberg (2005, 2006) obtained promising results from acoustic and prosodic information. Maskey and Hirschberg (2005) showed that the best performance was obtained by combining acoustic features with lexical, structural, and discursive ones in summarizing English broadcast news. They also demonstrated that acoustic and structural features perform well when the speech transcript is not available. Maskey and Hirschberg (2006) suggested that acoustic features are useful in extracting salient sentences in synopses of English broadcast news. Similarly, Zhang and Fung (2007) showed that prosodic and structural features permit the summarization of Chinese (Mandarin) broadcast news even without lexical features.

In a comprehensive study of the acoustic/prosodic, linguistic, and structural features of Chinese speech summarization, Zhang, Chan,

Fung, and Cao (2007b) suggested that the effectiveness of speech summarization depends on speech genres. According to their study, lexical features (e.g., word frequency) are more important for lecture speech summarization than acoustic features (e.g., pitch), which might be used to identify topic shifts or acoustically significant segments (Maskey, 2008). On the contrary, acoustic features are more important than lexical features for broadcast news.

Similarly, Zhang, Chan, and Fung (2007a) revealed that lexical features contribute to Chinese lecture summarization more than acoustic features. Further, they suggested that rhetorical structures (consisting of introductory, content, and concluding sections) occur in spoken lecture documents, and that acoustic and prosodic features permit the modeling of this rhetorical information to improve summarization performance. The normalization of acoustic features consistently improves summarization performance.

Indeed, Xie, Hakkani–Tur, Favre, and Liu (2009) showed that in creating summaries of meetings conducted in English, the employment of properly normalized acoustic features produces a performance comparable with or better than the use of lexical features. Similarly, Zhang and Fung (2012) showed that speaker–normalized acoustic features can improve the performance of Chinese lecture speech summarization.

6.1.2 Unsupervised Summarization

Unsupervised techniques do not, unlike supervised techniques, rely on annotated data, but on the linguistic and statistical information obtained from the document. Murray, Renals, and Carletta (2005) compared LSA, the MMR model, and feature-based summarization of a meeting corpus. They found that LSA, which represents each sentence of a document as a vector in its latent semantic space, recorded the best performance, while the MMR model did very well because of its ability to reduce redundancy.

On the contrary, Fujii et al. (2008) drew the opposite conclusion, finding that a feature-based method using linguistic and prosodic structures outperformed the MMR model in summarizing Japanese lecture speech data. They claimed that this was largely due to the difference in domains: lectures are usually organized using slides, thus containing less redundant information than recordings of meetings.

Meanwhile, Ribeiro and de Matos (2007) found that the MMR model is superior to both the feature-based and LSA approaches for Portuguese broadcast news. On the whole, most research on speech summarization has focused on extractive generic summarization of a single document, although some studies have relied on multiple documents. For example, Zhu, Penn, and Rudzicz (2009) used the recurrence statistics of acoustic patterns in multiple recordings without transcripts and the MMR model to identify salient utterances, measure the similarity between utterances (sentences), and to remove redundancy in speech summarization. Chen et al. (2014) proposed a recurrent neural network language modeling (RNNLM) framework, and they suggested that their proposed LM framework shows competitive results when compared to the other unsupervised methods.

6.1.3 Social Summarization

Video tags known to be similar to or better than indexer—assigned terms (Heckner, Neubauer & Wolff, 2008; Kim, 2011a) can be effectively used as key concepts to select significant sentences from speech transcripts. Heckner et al. (2008) claim that videos on YouTube are extensively tagged because uploaders want their videos to be retrieved and viewed by as many people as possible, and not because they want to organize their personal collection. Similarly, Kim (2011a) notes that video tags seem to be closer to indexer—assigned terms than image tags because YouTube videos have a proportionally greater number of tags describing video content than Flickr images.

More importantly, by using an acoustic pattern for a given tag, we can extract the utterance(s) containing the tag directly from the original speech to form a spoken summary in the absence of a transcript. Moreover, the function of such a tag method is similar to named−entity extraction, which utilizes the frequency of proper names of persons, locations, and organizations. Frequently assigned video tags are related to named entities, such as people and object categories (Kim, 2011a). Named entities, which are often related to topic−oriented words, are more important in speech summarization than in text summarization (Zhang, Cohn, & Ciravegna, 2013; Christensen, Gotoh, Kolluru, & Renalset, 2003).

Yamamoto, Masuda, Ohira, and Nagao (2008) proposed a video scene annotation method based on tag clouds. Their proposed system, Synvie, enables users to click on a tag in a tag cloud (generated from user comments) while watching a video, so that the clicked tag becomes associated with that particular time point of the video. They suggested that the proposed method resolved the problem of low coverage of tags faced in previous studies, and users were motivated to add tags by using the tag sharing and tag cloud methods.

Having statistically analyzing video bookmarks to form video storyboards, Chung, Wang, and Sheu (2011) extracted some meaningful key frames. In doing so, they assumed that the video frames around bookmarks added by users were sufficiently represent−ative for video summarization. Their proposed method produced better summaries than two existing methods that use audio−visual features. Hannon, McCarthy, Lynch, and Smyth (2011) proposed and evaluated two methods based on the frequency and content of time–stamped Twitter messages for offline generation of video highlights of the World Cup. The results indicated that test users were largely satisfied with the summaries produced by both techniques.

Wang et al. (2012) presented an effective method for summarizing the content of YouTube video search results by tag localization and key–shot mining. They first localized the user–assigned tags associated with each video to its specific scenes, and then identified a set of key shots based on relevance scores computed by matching shot–level tags with an event query. The key shots were then used for either a threaded video skim or a visual–textual storyboard.

Table 6-1. Selected studies on speech summarization

Study	Supervised				Unsupervised (MMR, LSA, LM)	Social summarization	Genre	Language	Result
	Le	St	Ac	Di					
Maskey & Hirschberg (2005)	Le	St	Ac	Di			broadcast news	English	− Le + St + Ac + Di: best performance − St + Ac: good performance without transcript

Study	Supervised				Unsupervised (MMR, LSA, LM)	Social summarization	Genre	Language	Result
	Le	St	Ac	Di					
Maskey & Hirschberg (2006)		St	Ac				broadcast news	English	− Ac: good performance for broadcast news without Le
Zhang & Fung (2007)	Le	St	Ac				broadcast news	Chinese	− Ac + St: good performance without Le
Zhang et al. (2007a)	Le	St	Ac	Di			lecture	Chinese	− Le: better performance than Ac
Zhang et al. (2007b)	Le	St	Ac				broadcast news and lecture	Chinese	− Ac + St: good performance for broadcast news − Le: good performance for lecture
Xie et al. (2009)	Le	St	Ac	Di			meeting	English	− normalized Ac: produces as good or even better performance than Le
Zhang & Fung (2012)	Le		Ac				lecture	Chinese	− speaker-normalized Ac: improves performance

Study	Supervised				Unsupervised (MMR, LSA, LM)	Social summarization	Genre	Language	Result
	Le	St	Ac	Di					
Murray et al. (2005)	Le		Ac		MMR, LSA		meeting	English	− LSA: best performance − MMR: comparable performance to LSA
Ribeiro & de Matos (2007)	Le	St			MMR, LSA		broadcast news	Portuguese	− MMR: best performance
Fujii et al. (2008)	Le		Ac		MMR		lecture	Japanese	− Le / Ac: better performance than MMR
Zhu et al. (2009)			Ac		MMR		broadcast news	English	− Ac can be used to measure the similarity between sentences
Chen et al. (2014)					LM		broadcast news	Chinese	− LM shows competitive results when compared to the other unsupervised methods
Yamamoto et al. (2008)						Co & Ta	unspecified	Japanese	− the coverage of annotations generated by the proposed method is higher than that of existing methods

Study	Supervised				Unsupervised (MMR, LSA, LM)	Social summarization	Genre	Language	Result
	Le	St	Ac	Di					
Chung et al. (2011)						Bo	unspecified	unspecified	– proposed method produced semantically more important summaries than existing methods
Hannon et al. (2011)						Tw	sports highlights	unspecified	– proposed two methods are satisfied by test users
Wang et al. (2012)						Ta	unspecified	English	– proposed approach is effective
Kim & Kim (2016)						Ta		English	

Bo: Bookmark; Tw: Twitter; Co: Comment; Ta: Tag Ac: Acoustic; St: Structural; Le: Lexical; Di: Discourse

6.2 Social Summarization

Social summarization can be defined as a summarization method that creates text or multimedia summaries by exploiting user feedback, such as tags, comments, notes, and bookmarks obtained from social networking (Kim & Kim, 2016). In this study (Kim, 2013), we proposed a tag–based framework to adapt the summ–

arization process of human abstractors in which important sentences are selected based on the keywords in a sentence and their semantic relations (Heu et al., 2013). This framework is designed to select sentences based on tags as key concepts as well as their semantic relations. Both WordNet 2.1 synonyms and Flickr tag clusters are used to expand tags, and to detect the semantic relations between them.

6.2.1 Tag-based Framework

We need to construct a specific model for speech summarization that utilizes social tags, expanded tags, and their semantic relations. For this purpose, we proposed the summarization model for assigning weights to each sentence of a document. In extractive summarization, the weight score of a sentence is calculated as follows:

$$W (S_i) = (Sim (S_i, T_{ex}) + Sem (S_i)) / 2$$

where T_{ex} is the final tag set obtained from a tag list (see Table 6-2) and Sim (S_i, T_{ex}) is a metric for computing the similarity of a sentence S_i and the expanded tag set T_{ex} using cosine similarity. To measure the relevance score between each sentence and a document, in addition to Sim (S_i, T_{ex}) we use Sem (S_i) to adjust the weight of the semantic relation between tags, which are obtained from Flickr tag clusters and WordNet synonyms. We then take the average of Sim (S_i, T_{ex}) and Sem (S_i). Last, sentences with average scores greater than a given threshold were extracted to form a summary.

6.2.2 The Process of Generating Tag-based Summaries .

This section describes the three−step process of generating social speech summaries by the tag−based framework (Figure 6−1). After collecting sample videos, the three steps (sentence segmentation, tag expansion, and sentence score computation) are related to extracting

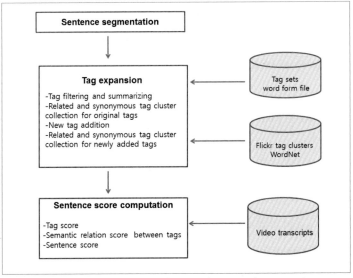

Figure 6-1. The process of generating speech summaries

speech summaries. That is, we collected a transcript for each of the 30 videos that were sampled from the Ted Talk and YouTube sites. We then segmented a transcript into sentences. Next, we assigned a score to each sentence through the following two tasks (tag expansion and sentence score computation).

1) Tag expansion

(1) Tag filtering and summarizing: As shown in Table 6–2, of the 14 original tags associated with Video 11, six that do not appear in its transcript were deleted. An initial tag set was formed that includes eight single–term tags (T_1–T_8 [Stanford, commencement, apple, graduation, NeXt, Pixar, cancer, & computer]).

(2) Related and synonymous tag cluster collection for original tags: We first collected tag clusters for each of the eight tags in the initial tag set using Flickr. The second tag, "commencement" has two tag clusters (tag–cluster 1: graduation, university, college; tag–cluster 2: Tacoma, bay, Washington). We select a tag–cluster if it includes the same tag(s) as in the tag set. Of the two tag–clusters, we selected tag–cluster 1 because it contains the word "graduation," which belongs to the tag set.

Next, we collected tag clusters for each of the eight tags

from the initial tag set using WordNet. The second tag, "commencement," has three senses (clusters) (sense 1: beginning, first, outset···; sense 2: commencement exercise, commencement ceremony, graduation, graduation exercise; sense 3: beginning, start). We selected sense 2 because it contains the word "graduation," which belongs to the tag set. Thus, tag–cluster 1 and sense 2 are stored as TC_2 in Table 6–2. Tag cluster collection was conducted for the remaining tags using the same procedure.

(3) New tag addition: Out of the ten words collected from the three tag–clusters (TC_2, TC_4, TC_8), we added only two new tags (university and college). Five words (commencement exercise, commencement ceremony, graduation exercise, mac, & laptop) were not selected because they were not included in the transcript, and three words (graduation, commencement, & Apple) were not selected because they were already included in the initial tag set.

(4) Related and synonymous tag cluster collection for newly added tags: According to the aforementioned procedure, we selected tag–clusters for each of the two new tags. As a result, we had two new tags ("campus" and "school") that appear in the transcript. However, we did not add them to the tag list in the second tag–cluster collection. Eventually,

we obtained a final tag list that consisted of ten single-term tags. A weight of 1 was assigned to each one-word tag and 2 to each compound tag. A tag weight table (TWT) can be represented as follows:

$$\text{TWT} = \{(t_1, w_1), (t_2, w_2), \cdots, (t_{n-1}, w_{n-1}), (t_n, w_n)\}$$

where t_i is the i-th tag in a document and w_i is its weight.

2) Sentence score computation

(1) Tag score: To compute the tag score ($\text{Sim}_1 (S_1, T_{ex})$) of sentence 1 ($S_1$) ("I am honored to be with you today at your commencement from one of the finest university in the world") in Video 11, we created the following tag (T_{ex}) and S_1 vectors using the ten tags listed in Table 6-2:

$$T_{ex} = (1.0, 1.0, 1.0, 1.0, 1.0, 1.0, 1.0, 1.0, 1.0, 1.0)$$
$$S_1 = (0.0, 1.0, 0.0, 0.0, 0.0, 0.0, 0.0, 0.0, 1.0, 0.0)$$

The tag score of S_1, which includes two tags (commencement and university), is 0.45.

(2) Semantic relation score between tags: To compute the semantic relation score between tags, we adapted the algorithm used by Heu et al. (2013). To calculate Sem (S_1),

we first identified the relation among tags using a tag relation table (TRT) (see Table 6-3) that was constructed using Table 6-2. For example, as shown in Figure 6-2, TC_2 includes T_4 and TC_4 includes T_2. Thus, T_2 and T_4 have a strong semantic relation, because their tag–clusters include both tags. T_2 and T_9 have a weak semantic relation, because only TC_2 includes T_9, and T_1 and T_2 have no relation because their tag–clusters do not include the other's tag.

Table 6-2. Tag list for Video 11

	Tag (=T_i)	Tag cluster (=TC_i)	
		Flickr related tags	WordNet synonyms
1	Stanford	–	–
2	commencement	graduation, university, college	commencement exercise[*], commencement ceremony[*], graduation, graduation exercise[*]
3	Apple	–	–
4	graduation	university, college, commencement	–
5	NeXT	–	–
6	Pixar	–	–
7	cancer	–	–
8	computer	Apple, mac[*], laptop[*]	–
9	**university**	college, campus, architecture[*]	–
10	**college**	university, campus, school	–

Tags underlined were already included in the initial tag list; Tags and synonyms with superscript asterisks were not included in the transcript of Video 11; two tags (in bold fonts) were added as new tags.

Table 6-3. Tag Relation Table (TRT(D_{11})) for Video 11

	T_1	T_2	T_3	T_4	T_5	T_6	T_7	T_8	T_9	T_{10}
TC_1										
TC_2				0					0	0
TC_3										
TC_4		0							0	0
TC_5										
TC_6										
TC_7										
TC_8			0							
TC_9										0
TC_{10}									0	

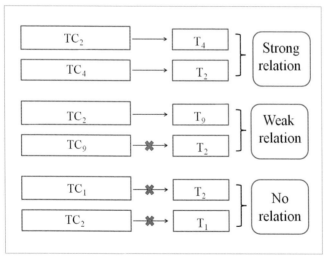

Figure 6-2. Semantic relations between tags

We calculate the semantic relationship between tags in a sentence

(S_i) as follows:

$$Sem\ (S_i) = \Sigma\ R_{i,j}$$
$$i, j \in trt(s_{i_j}),\ trt(s_{i_j}) \subseteq TRT(D_i)$$

where $R_{i,j} = 0.1$ if $T_j \in TC_i$ and $R_{i,j} = 0$ if $T_j \notin TC_i$. $TRT(D_i)$ is the tag relation table for document D_i, and $trt(s_i)$, a subset of $TRT(D_i)$, is the tag relation table for sentence S_i that belongs to document D_i.

For example, to compute the semantic relation score ($Sem\ (S_1)$) between the tags in sentence 1 (S_1) of Video 11, we created $trt(S_1)$ (see Table 6–4), a subset of $TRT(D_{11})$; S_1 includes two tags (commencement [T_2] and university [T_9]). After substituting the tag relation values obtained from $trt(S_1)$ into the formula for $Sem\ (S_i)$, we found that $Sem\ (S_1) = 0.1\ (=[R_{2,9}])$, because only one tag pair (T_2 and T_9) have a weak semantic relation.

Table 6-4. tag relation table (trt(S_1)) for S_1 of Video 11

	T_2	T_9
TC_2	–	0
TC_9	–	–

(3) Sentence score: To obtain a sentence score, we first

computed the average value of Sim (S_i, T_{ex}) and Sem (S_i) in the summarization model. For example, the Sim (S_1, T_{ex}) score of Video 11 was 0.45 and its Sem (S_1) score was 0.10, so its sentence score was 0.28 ([0.45+0.10] / 2). As a result, the final summary of Video 11 consists of six sentences with average scores greater than 0.21.

Most speech transcripts were summarized within the compression rate range of 3~10%. We tried to limit the number of sentences in each summary to between 2 and 8, and the duration of each summary to between 15 and 70 s, because we considered it difficult to maintain focus when listening to a long spoken summary.

6.3 Latent Semantic Analysis (LSA)

6.3.1 Overview

LSA is an extended model of the classic vector space model proposed by Salton, Wong, and Yang (1975). It is used to identify latent patterns in the complex relationships between a set of documents (sentences) and the words constituting them. This is carried out by generating a set of concepts related to the documents and the words (Yi, 2011).

LSA uses the context of an input document (word–document matrix). Furthermore, it retains information about words used in a sentence and the common words found in different sentences. A large number of words in common between sentences implies that the sentences are highly semantically related to each other (Ozsoy, Cicekli, & Alpaslan, 2010; Ozsoy, Alpaslan, & Cicekli, 2011). The meaning of a sentence is determined by using the words that it contains, and meanings of words are decided by using the sentences that contains the words. LSA applies singular value decomposition (SVD) to derive a latent semantic structure for sentences in a document by clustering content words.

6.3.2 The Process of Generating LSA-based Summaries

We constructed the LSA–based summaries using the following procedure (Kim & Kim, 2016):

1) Using Wordcounter (http://www.wordcounter.com/), we obtained a list of words that occurred at least once in the transcript of each sample video. We then excluded stop words (e.g., "the," "it," etc.) from the word list.

2) Using each word list, we created two input word–sentence matrices to examine the effect of matrix size on the

effectiveness of summarization. The first matrix was created with a list of words with a frequency of at least two. The second matrix was created using a different procedure, where words with more than two occurrences were first selected, and content words were chosen from the initially selected words to form a word list. As a result, the average number of words in the first matrix was 81.4, and that in the second matrix was 12. The LSA method using the first matrix is signified as the LSA (avg. = 81.4) method, whereas the LSA method using the second matrix is referred to the LSA (avg. = 12.1) method.

3) To semantically cluster keywords and derive a latent semantic structure for sentences in a document, we utilized MATLAB's svds (A, k) function, which computes the k largest singular values and the associated singular vectors of matrix A. In our study, k was assigned the value 2. Further, to select important sentences from a V^T matrix (see Table 6−5), we used a cross method because it performs better than other LSA−based approaches (Ozsoy et al., 2011).

The cross method is one of various approaches for sentence selection while creating summaries using the LSA, and was proposed by Ozsoy et al. (2010). The cross method uses a pre−processing step between SVD calculation and sentence selection, which enables to remove the overall effect of sentences that are partially, but not mainly related to a concept (topic). That is, for each concept represented by the rows of the V^T matrix, the average sentence score was calculated. Cell values less than or equal to the average score were set to zero. The total length of each sentence vector was then calculated, and the longest sentence vectors were collected for a summary.

Table 6-5. Example of cross method (Kim and Kim, 2016)

	V^T matrix (k = 2)			
	Sent1	Sent2	Sent3	Avg.
Con1	0.46 → 0	0.73	0.50 → 0	0.56
Con2	−0.67 → 0	0.04	0.54	−0.03
Length	0	0.77	0.54	

6.4 Maximum Marginal Relevance (MMR)

MMR is designed to generate more diverse summaries by choosing the most relevant sentences, while ensuring that there is minimal duplication of sentences already chosen for a proposed summary (Provost, 2008). In the study (Kim, 2011c), we proposed a MMR model, and then explained the process of generating social speech summaries by using the MMR model.

6.4.1 MMR Model

The MMR model is designed to choose the most relevant remaining sentence, while ensuring that there is minimal duplication of any sentences already chosen for a proposed summary, which leads to the generation of more diverse summaries (Provast, 2008). In the extractive summarization, the score of a sentence S_i in the kth iteration of the MMR is calculated as follows:

$$MMR = \underset{S_i \in R \backslash S}{arg\ max} \left[\lambda\ Sim_1\ (S_i,\ T) - (1 - \lambda)\ \underset{S_j \in S}{max}\ Sim_2\ (S_i,\ S_j) \right]$$

where T is a tag set, R is the set of all sentences in the document set, S is the current set of already selected sentences for the summary, and $R \backslash S$ is the set of unselected sentences in R. Sim_1 is the similarity metric for computing the similarity between a sentence S_i and the tag set T, and Sim_2 is the similarity metric for computing the similarity between two sentences S_i and S_j. λ is a parameter used to adjust the combined score in order to emphasize relevance or to avoid redundancy (Liu & Hakkani−Tur, 2011); we set the value of λ as 0.6. The sentences with the highest MMR scores will be iteratively chosen in the summary until it reaches a predefined proper size.

6.4.2 The Process of Generating MMR-based Summaries

The following section describes the process of social summary construction in three steps (tag filtering and summarization, tag score computing, and MMR score computing).

(1) Tag Filtering and Summarization: We collected a tag set for each of 8 test videos form the YouTube site. Then from each tag list, we removed spam tag(s) and summarized tags with a word-form file, which is used to collect tags with the same meaning, since we had to collect as many tags with the same meaning as possible. The word-form file was created to summarize diverse compound, singular, and plural forms, abbreviations and full-names, and different languages. Table 6-6 shows a final tag list for Video 3 (whose title is "What I do for Open Government") with 5 compound tags and 13 single term tags of eight sample videos, after filtering and grouping tags; there is no spam tag and "sphere/spheres" and "blog/blogging" are summarized, respectively. Table 6-6 shows 3 compound tags and 6 single term tags that occur in the transcript of Video 3 and 9 other tags ("Kate Lundy," "Web 2.0," "2.0," "transparency," "politics," "politician," "Canberra," "open," and "wiki") that did not appear in the transcript of Video 3. We assigned a weight value of "1" to

each of the 3 compound tags and a weight value of "0.5" to each of the 6 one-word tags.

Table 6-6. A tag list of Video 3

Tag No.	Tag	Weight
1	government 2.0	1
2	open government	1
3	public sphere	1
4	Australia	0.5
5	blog(ging)	0.5
6	government	0.5
7	public	0.5
8	sphere(s)	0.5
9	twitter	0.5

(2) Tag Score Computing: A tag score (Sim_1) was computed from the match between tags and words in a sentence. To do so, we collected a transcript for each of the 8 test videos from the YouTube site and segmented each transcript into sentences using period marks. Subsequently, a tag score was counted using the cosine similarity for each sentence of a given transcript. For example, using the tag list (Table 6-6), we selected 10 candidate sentences whose tag scores are equal or greater than 0.33 from 25 sentences of the transcript of Video 3 (see Figure 6-3).

(3) MMR Score Computing: In order to obtain MMR scores, we needed the similarity values between sentences (Sim_2), as well as the tag scores (Sim_1) obtained in the previous step. We also used the cosine similarity to get the values of Sim_2. To do so, we employed the RankWords program to extract a list of words used in the transcript of Video 3 and ranked them according to their frequency so as to obtain 21 nouns or proper words whose frequencies were greater than one (see Table 6-7); we assigned a weight value of "1.0" to a one-word tag, a weight value of "2.0" to a compound tag, and a weight value of "0.5" to the remaining words.

Table 6-7. A word list of Video 3

Word No.	Word	Weight
1	Australia	1.0
2	blog(ging)	1.0
3	channels	0.5
4	citizen(s)	0.5
5	commonwealth	0.5
6	conversation(s)	0.5
7	environment	0.5
8	government	1.0
9	government 2.0	2.0
10	House	0.5
11	information	0.5
12	initiatives	0.5

Word No.	Word	Weight
13	networking(ing)	0.5
14	open government	2.0
15	public	1.0
16	public sphere	2.0
17	site	0.5
18	sphere(s)	1.0
19	technology	0.5
20	twitter	1.0
21	White	0.5

Next, in order to measure MMR scores, we first selected the sentence (S20) with the highest tag score and then measured the similarity between S20 and each of the 9 unselected sentences (Sim$_2$). For example, the cosine similarity ("0.0") between S20 and S4 was obtained using the following two sentence vectors:

$$S20 = (0,0,0,0,\ 0,0,0,0,\ 2,0,0,0,\ 0,0,0,2,0,0.0,0,0)$$
$$S4\ = (0,0,0,0,\ 0,0,0,2,\ 0,0,0,0,\ 0,2,0,0,0,0.0,0,0)$$

Finally, we computed the MMR score (0.32) with the values of both Sim$_1$ (0.53) and Sim$_2$ (0.0). Figure 6-3 indicates that S4 had the highest MMR score (0.32) (0.53 multiplied by 0.6) and was selected. This procedure was repeated until the summary reached a predefined proper size (e.g., 5 sentences for Video 3).

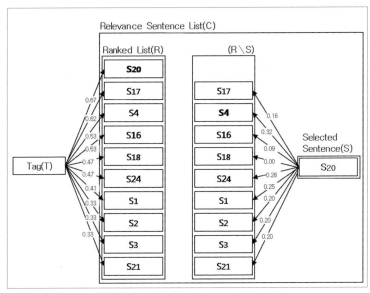

Figure 6-3. MMR scores for Video 3

6.5 Acoustic Method

6.5.1 Overview

Supervised techniques use lexical, structural, and discourse attributes. Many studies suggested that acoustic features show competitive results when compared to the other features (Maskey & Hirschberg, 2005; Zhang & Fung, 2012). Three acoustic factors, speaking rate (number of voiced frames/total frames), pitch pattern, and intensity (or subband energy) are favored by many researchers

(Maskey, 2008; Hirschberg, 2002; Zhang & Fung, 2012) for tagging the important contents of speeches. Assuming that speakers may talk more loudly if they want to emphasize significant segments of speech, we utilized intensity. Similarly, assuming that topic shifts may be marked by changes in pitch (Hirschberg & Nakatani, 1996) and that new topics are often introduced with content-laden sentences, which are often included in a summary (Maskey, 2008), we regarded pitch patterns. Additionally, speaking rate is employed, under the supposition that speakers may make sudden changes in speaking rates when emphasizing significant segments of speech.

In the studies (Kim, 2011a; Kim, 2012c), we first investigated whether these features are equally important and, if not, which one can be effectively modeled to compute the significance of segments for lecture summarization. Thus we extracted, using Praat 5.2.42, the speaking rate, pitch (F0 maximum [max], F0 minimum [min], F0 mean, and difference [Diff] between F0 max and F0 min), and intensity (max DB, min DB, mean DB, and the Diff between max DB and min DB) features of each audio file of the 40 sample videos. We found that among the nine features, four (F0 max, the Diff between F0 max and F0 min, min DB, and the Diff between max DB and min DB) are discriminatory, while the remaining five are not; for example, in most sample videos, the speaking rates of all segments of a spoken document tend to be the same.

Of the four discriminatory acoustic features, the F measures (0.17 [Diff between max DB and min DB], 0.15 [min DB]) of intensity are higher than those (0.10 [F0 max], 0.09 [Diff between F0 max and F0 min]) of pitch, which confirms the result of Wang and Narayanan (2007) that intensity is the most useful feature for discriminating word prominence in speech.

6.5.2 The Process of Generating Acoustic-based Summaries

We explain the process of generating acoustic−based speech summaries (Kim, 2012c) using Praat (http://www.fon.hum.uva.nl/praat/). The process has three steps:

1) First Step: We extracted four (F0 max, the Diff between F0 max and F0 min, min DB, and the Diff between max DB and min DB) features, using Praat 5.2.42 (Figure 6−4).

2) Second Step: Relying on the four discriminatory features, we made 40 speech summaries. During this process, we normalized each feature category in a sentence by using the maximum, minimum and range values of pitch and intensity of a spoken document; thus each feature category is within the range of [0, 1] (Figure 6−5 and 6−6).

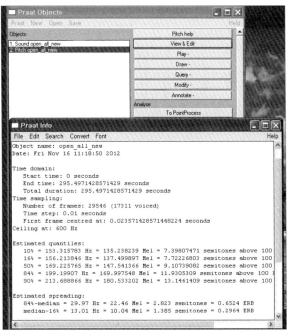

Figure 6-4. Praat interface

3) Third Step: Speech summaries automatically generated in the previous step can be presented in either speech or text form.

We first investigated whether acoustic features (speaking rate, pitch pattern, and intensity) are equally important and, if not, which one can be effectively modeled to compute the significance of segments for lecture summarization. As a result, we found that the intensity (that is, difference between max DB and min DB) is the most efficient factor for speech summarization.

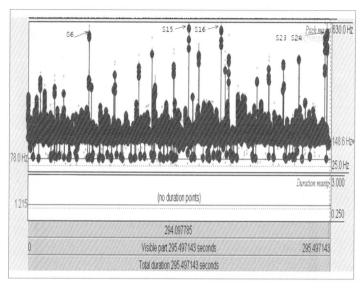

Figure 6-5. Pitch analysis (Video 3)

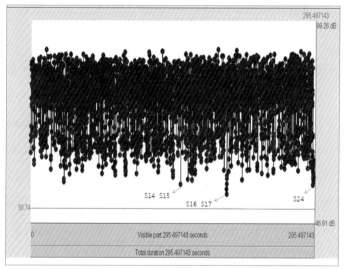

Figure 6-6. Intensity analysis (Video 3)

References

Abdi, A., Idris, N., Alguliyev, R. M., & Aliguliyev, R. M. (2017). Query−based multi−documents summarization using linguistic knowledge and content word expansion. Soft Computing, 21(7), 1785~1801.

Cawkell, A. (1995). A guide to image processing and picture management. Aldershot, Hampshire: Gower Publishing Ltd.

Chen, B. & Lin, S. (2012). A risk−aware modeling framework for speech summarization. IEEE Transactions on Audio, Speech, and Language Processing, 20(1), 211~222.

Chen, K. Y., Liu, S. H., Chen, B., Wang, H. M., Hsu, W. L., & Chen, H. H. (2014, July). A recurrent neural network language modeling framework for extractive speech summarization. In Multimedia and Expo (ICME), 2014 IEEE International Conference on (pp. 1~6). IEEE.

Chowdhury, G. G. (2010). Introduction to modern information retrieval. Facet Publishing.

Christensen, H., Gotoh, Y, Kolluru, B, & Renalset, S. (2003). Are extractive text summarisation techniques portable to broadcast news? In Proceedings of Automatic Speech Recognition and Understanding Workshop (pp. 489~494). St. Thomas, USA.

Chung, M., Wang, T., & Sheu, P. (2011). Video summarisation based on collaborative temporal tags. Online Information Review, 35(4), 653~668.

CumbleyAuthor Vitae, R. & Church, P. (2013). Is "big data" creepy?

Computer Law & Security Review, 29(5), 601~609.

Ding, W., Marchionini, G., & Soergel, D. (1999). Multimodal surrogates for video browsing. In Proceedings of the Fourth ACM Conference on Digital Libraries (pp. 85~93). Berkeley, CA.

Fujii, Y., Yamamoto, K., Kitaoka, N. & Nakagawa, S. (2008). Class lecture summarization taking into account consecutiveness of important sentences. In Proceedings of Interspeech (pp. 2438~2441).

Furui, S., Kikuchi, T., Shinnaka, Y., & Hori, C. (2004). Speech-to-text and speech-to-speech summarization of spontaneous speech. IEEE Trans. Speech Audio Process, 12(4), 401~408.

Goldstein, J. Mittal, V., Carbonell, J., & Kantrowitz, M. (2000). Multi-document summarization by sentence extraction. In Proceedings of the 2000 NAACL-ANLP Workshop on Automatic Summarization (Vol. 4., pp. 40~48).

Gong, Y., & Liu, X. (2001, September). Generic text summarization using relevance measure and latent semantic analysis. In Proceedings of the 24th Annual International ACM SIGIR Conference on Research and Development in Information Retrieval (pp. 19~25). ACM.

Hannon, J., McCarthy, K., Lynch, J., & Smyth, B. (2011). Personalized and automatic social summarization of events in video. In Proceedings of the 16th International Conference on Intelligent User Interfaces (pp. 335~338). Palo Alto, California, USA.

Heckner, M., Neubauer,T., & Wolff, C. (2008). Tree, funny, to_read, Google: What are tags supposed to achieve? In Proceedings of the 2008 ACM Workshop on Search in Social Media (pp. 3~10). New

York: ACM Press.

Heu, J. et al. (2013). Multi–document summarization exploiting semantic analysis based on tag cluster. In S. Li et al. (eds.), Advances in multimedia modeling, Lecture Notes in Computer Science, 7733, 479~489.

Hirohata, M., Shinnaka, Y., Iwano, K., & Furui, S. (2006). Sentence–extractive automatic speech summarization and evaluation techniques. Speech Communication, 48(9), 1151~1161.

Hirschberg, J. (2002). Communication and prosody: Functional aspects of prosody. Speech Communication, 36(1–2), 31~43.

Hirschberg, J., & Nakatani, C. (1996). A prosodic analysis of discourse segments in direction–given monologues. In Proceedings of the 34th Annual Meeting of the Association for Computational Linguistics (pp. 286~293). Santa Cruz, California.

Iyer, H., & Lewis, C.D. (2007). Prioritization strategies for video storyboard keyframes. Journal of the American Society for Information Science and Technology, 58(5), 629~644.

Jorge–Botana, G., León, J., Olmos, R., & Hassan–Montero, Y. (2010). Visualizing polysemy using LSA and the predication algorithm. Journal of the American Society for Information Science and Technology, 61(8), 1706~1724.

Jung, H. T., Kim, D. W., Kim, S., Im, C. H., & Lee, S. H. (2012). Reduced source activity of event–related potentials for affective facial pictures in schizophrenia patients. Schizophrenia Research, 136(1), 150~159.

Kim, H., & Kim, Y. (2010). Toward a conceptual framework of key–frame extraction and storyboard display for video summarization. Journal of the American Society for Information Science and Technology, 61(5), 927~939.

Kim, H. (2011a). Toward video semantic search based on a structured folksonomy. Journal of the American Society for Information Science, 62(3), 478~492.

Kim, H. (2011b). A study on the interactive effect of spoken words and imagery not synchronized in multimedia surrogates for video gisting. Journal of the Korean Society for Library and Information Science, 45(2), 97~118.

Kim, H. (2011c), Social speech summarization. Proceedings of the Association for Information Science and Technology, 48(1), 1~4.

Kim, H. (2012a). A tag-based framework for extracting spoken surrogates. Proceedings of the ASIST Annual Meeting, 49. Medford, NJ: Information Today.

Kim, H. (2012b). Investigating an automatic method in summarizing a video speech using user-assigned tags. Journal of the Korean society for Library and Information Science, 46(1), 163~181.

Kim, H. (2012c). Investigating an automatic method for summarizing and presenting a video speech using acoustic features. Journal of Korean Society for Information Management, 29(4), 191~208.

Kim, H. (2013). Comparing the use of semantic relations between tags versus latent semantic analysis for generic speech summarization. Proceedings of the ASIST Annual Meeting, 50. Medford, NJ: Information Today.

Kim, H., & Kim, Y. (2016). Generic speech summarization of transcribed lecture videos: Using tags and their semantic relations. Journal of the Association for Information Science and Technology, 67(2), 366~379.

Lin, S., Chen, B., & Wang, H. (2009). A comparative study of probabilistic ranking models for Chinese spoken document summarization. ACM Transactions on Asian Language Information Processing, 8(1), 3:1~3:23.

Liu, Y., & Hakkani-Tür, D. (2011). Speech summarization. In G. Tur & R. De Mori (Eds.), Spoken language understanding: Systems for extracting semantic information from speech (pp. 357~392). Chichester, UK: John Wiley & Sons, Ltd.

Maskey, S. (2008). Automatic broadcast news speech summarization. Unpublished doctoral dissertation, Columbia University.

Maskey, S., & Hirschberg, J. (2005). Comparing lexical, acoustic/prosodic, structural and discourse features for speech summarization. In Proceedings of Interspeech (pp. 621~624).

Maskey, S., & Hirschberg, J. (2006). Summarizing speech without text using Hidden Markov Models. In Proceedings of the Human Language Technology Conference of the NAACL (Companion Volume: Short Papers, pp. 89~92). Association for Computational Linguistics, Stroudsburg, PA, USA.

Marchionini, G., Song, Y., & Farrell, R. (2009). Multimedia surrogates for video gisting: Toward combining spoken words and imagery. Information Processing and Management, 45(6), 615~630.

Mendoza, M., Bonilla, S., Noguera, C., Cobos, C., & León, E. (2014). Extractive single-document summarization based on genetic operators and guided local search. Expert Systems with Applications, 41(9), 4158~4169.

Moshfeghi, Y., Pinto, L. R., Pollick, F. E., & Jose, J. M. (2013, March). Understanding relevance: An fMRI study. In P. Serdyukov et al., eds. Advances in information retrieval. Springer Berlin Heidelberg.

Murray, G., Renals, S., & Carletta, J. (2005). Extractive summarization of meeting recordings. In Proceedings of the 9th European Conference on Speech Communication and Technology (INTERSPEECH)(pp. 593~596). Lisbon, Portugal.

Myaeng. S., & Jang. D. (1999). Development and evaluation of a statistically−based document summarization system. In I. Mani and M. T. Maybury (eds.), Advanced in automatic text summarization (pp. 61~70). Cambridge. Massachusetts: the MIT Press.

Naghavi, H., & Nyberg, L. (2005). Common fronto−parietal activity in attention, memory, and consciousness: Shared demands on integr−ation? Consciousness and Cognition, 14(2), 390~425.

Ozsoy, M., Alpaslan, F., & Cicekli, I. (2011). Text summarization using latent semantic analysis. Journal of Information Science, 37(4), 405~417.

Ozsoy, M., Cicekli, I., & Alpaslan, F. (2010). Text summarization of Turkish texts using latent semantic analysis. In Proceedings of the 23rd International Conference on Computational Linguistics (pp. 869~876).

Park, J. et al. (2008). Web content summarization using social bookmarks: A new approach for social summarization. In Proceedings of the 10th

ACM Workshop on Web Information and Data Management (pp. 103~110).

Patil, K., & Brazdil, P. (2007). Sumgraph: Text summarization using centrality in the pathfinder network. International Journal on Computer Science and Information Systems, 2(1), 18~32.

Peters, I., & Stock, W.G. (2007). Folksonomy and information retrieval. Proceedings of American Society for Information Science and Technology Annual Meeting, 44(1), 1~18.

Provost, J. (2008). Improved document summarization and tag clouds via singular value decomposition. Unpublished master's thesis. Queen's University, Kingston, Ontario, Canada.

Quintarelli, E. (2005). Folksonomies: Power to the people. Paper presented at the ISKO Italy UniMIB meeting, Milan, Italy.

Ribeiro, R., & de Matos, D. (2007). Extractive summarization of broadcast news: Comparing strategies for European Portuguese. In V. Matousek & P. Mautner (Eds.), Text, speech and dialogue, Lecture Notes in Computer Science, 4629, 115~122.

Salton, G., Wong, A., & Yang, C. S. (1975). A vector space model for automatic indexing. Communications of the ACM, 18(11), 613~620.

Sato, T., et al. (1998). Video OCR for digital news archive. In Proceedings Workshop on Content-based Access of Image and Video Databases, 52~60, Los Alamitos, USA.

Smith, G. (2008). Tagging: People-powered metadata for the social Web. Berkeley: New Riders.

Specia, L., & Motta, E. (2007). Integrating folksonomies with the semantic

6. Speech Summarization 165

Web. In E. Franconi, M. Kifer, & W. May (Eds.), The semantic Web: Research and applications, Lecture Notes in Computer Science, 4519, 624~639.

Stanton, J. (2012). Big data and the library professional. Journal of the Library Administration & Management Section, 8(2), 22~24.

Thatcher, R. W., Krause, P. J., & Hrybyk, M. (1986). Cortico-cortical associations and EEG coherence: A two-compartmental model. Electroencephalography and Clinical Neurophysiology, 64(2), 123~143.

Togawa, H., & Okuda, M. (2005). Position-based keyframe selection for human motion animation. Proceedings of 11th International Conference on Parallel and Distributed Systems, Workshops (Vol. 2, pp. 182~185). Washington, DC: IEEE.

Tulving, E., Kapur, S., Craik, F. I., Moscovitch, M., & Houle, S. (1994). Hemispheric encoding/retrieval asymmetry in episodic memory: Positron emission tomography findings. Proceedings of the National Academy of Sciences, 91(6), 2016~2020.

Turner, J. (1994). Determining the subject content of still and moving documents for storage and retrieval: An experimental investigation. Unpublished doctoral dissertation, University of Toronto.

Turney, P. (2000). Learning algorithms for keyphrase extraction. Information Retrieval, 2(4), 303~336.

van Houten, Y., Oltmans, E., & van Setten, M. (2000). Video browsing and summarization (Rep. No. TI/RS/2000/63). Enschede: Telematica Instituut.

Wang, D., & Narayanan, S. (2007). An acoustic measure for word prominence in spontaneous speech. IEEE Transactions on Audio, Speech, and Language Processing, 15(2), 690~701.

Wang, M. et al. (2012). Event driven Web video summarization by tag localization and key-shot identification. IEEE Transactions on Multimedia, 14(4), 975~985.

Wang, S, Zhu, Y, Wu, G., & Ji, Q. (2013). Hybrid video emotional tagging using users' EEG and video content. Multimed Tools and Applications, 72(2), 1257~1283.

Wang, S., Zhu, Y., Wu, G., & Ji, Q. (2014). Hybrid video emotional tagging using users' EEG and video content. Multimedia Tools and Applications, 72(2), 1257~1283.

Xie, S., Hakkani-Tur, D., Favre, B., & Liu, Y. (2009). Integrating prosodic features in extractive meeting summarization. In Proceedings of the 11th Biannual IEEE Workshop on Automatic Speech Recognition and Understanding (pp. 387~391). Merano, Italy.

Yamamoto, D., Masuda, T., Ohira, S., & Nagao, K. (2008). Collaborative video scene annotation based on tag cloud. In Proceedings of the Advances in Multimedia Information Processing (pp. 397~406). Tainan, Taiwan.

Yi, K. (2011). An empirical study on the automatic resolution of semantic ambiguity in social tags. Proceedings of American Society for Information Science and Technology Annual Meeting, 48(1), 1~10.

Yokoi, K., Nakai, H., & Sato, T. (2008). Toshiba at TRECVID 2008: Surveillance event detection task. Proceedings of TRECVID 2008.

Zhang, J., Chan, H., & Fung, P. (2007a). Improving lecture speech summarization using rhetorical information. In Proceedings of the IEEE Workshop on Automatic Speech Recognition and Understanding (pp. 195~200). Kyoto, Japan.

Zhang, J., Chan, H., Fung, P., & Cao, L. (2007b). A comparative study on speech summarization of broadcast news and lecture speech. In Proceedings of the Annual Conference of the International Speech Communication Association (pp. 2781~2784). Antwerp, Belgium.

Zhang, J., & Fung, P. (2007). Speech summarization without lexical features for Mandarin broadcast news. In Proceedings of NAACL HLT (Companion Volume, pp. 213~216). Rochester, NY.

Zhang, J., & Fung, P. (2012). Active learning with semi-automatic annotation for extractive speech summarization. ACM Transactions on Speech and Language Processing, 8(4), 1~25.

Zhang, Z., Cohn, T., & Ciravegna, F. (2013, March). Topic-oriented words as features for named entity recognition. In International Conference on Intelligent Text Processing and Computational Linguistics (pp. 304~316). Springer, Berlin, Heidelberg.

Zhu, X., Penn, G., & Rudzicz, F. (2009). Summarizing multiple spoken documents: Finding evidence from untranscribed audio. In Proceedings of ACL / AFNLP (pp. 549~557). Suntec, Singapore.

Chapter 7

Video Summarization

7. Video Summarization

7.1 Overview

Video skims are one of ideal video summarization forms. However, it is very expensive to generate it. Thus, we use unsyn-chronized combination forms of image and audio surrogates (Figure 7-1). However, some users are irritated by such unsynchronized form while watching them.

Figure 7-1. Video storyboard

The storyboard is the most widely used visual (image) surrogate for videos and is a primary cause of two major problems pertaining to video data: identification of the optimal method of key–frame extraction from videos and the pattern employed in their subsequent display to an audience, in the form of a storyboard. The former problem has received attention in the field of computerized image processing with regard to the notions of shot–or scene–level boundary detection, whereas the latter problem has been addressed in the field of information science in studies on visual abstraction. At least three lines of approaches to key–frame extraction problem are available.

The first line of approach employs a computerized technique of shot boundary detection (or SBD), which automatically detects changes in formal features over shot boundaries. It is a trivial task to identify a representative key frame for each shot, exceeding a given duration threshold, between the detected shot boundaries (Smeaton & Browne, 2006; Dumont & Merialdo, 2007). This approach, despite the consistently high accuracy of its shot boundary detection (Smeaton, Over, & Doherty, 20010; Danisman & Alpkocak, 2006; Liu et al., 2007), also faces problems, including content redundancy between adjunct shots and issues stemming from the potentially large shot volume of a given video (e.g., a 2–min video may contain 24~60 distinct shots, which makes it very difficult to compose a coherent storyboard, as there are simply too many key frames to incorporate). Mundur, Rao, and Yesha (2006) proposed a video summarization technique that can eliminate redundant shots. Detyniecki and Marsala (2007) proposed an algorithm that identifies the longest shot from a set of similar adjunct shots and subsequently extracts a representative frame from it.

The second line of approach to key–frame extraction employs the content features of a video, such as motion (Togawa & Okuda, 2005), event (Yokoi, Nakai, & Sato, 2008), and overlay text (Sato et al., 1998; Anthimopoulos et al., 2007; Uchida et al., 2008). The content–based techniques for key–frame extraction have not yet been developed to the point where they are robust or effective in dealing with large collections (Smeaton et al., 2008).

The third line of approach employs a computerized technique of scene boundary detection. A single shot is content–bearing to some extent and is often sufficient for a user to gain a comprehensive understanding; however, in some cases, it is not. In many cases, content attributes such as persons, objects, activities, and settings span multiple shots, crossing shot boundaries. Such story elements lie at the scene level, composed of a series of shots that collectively describe the mood and the message. "It is the combination of shots what describes the story element, and each shot uses the context of the surrounding shots to convey its message." (Smeaton, 2007, p.557). This statement implies that the meaning of a shot is not typically determined within the shot's boundaries.

Many researchers have studied the concept of video scene partitioning, working under the assumption that inclusion of a scene that is composed of several shots, creatively combined together, enables viewers to acquire more meaningful content from a video. Previous studies have employed a number of different video attributes, including color features, audio-visual features, and low- and high-level features, for the purposes of scene boundary detection (Rasheed et al., 2005; Ariki et al., 2003; Zhu & Liu, 2009a, 2009b). Despite the improvement of the detection algorithms, the results of these studies remain unsatisfactory because of the gap that exists between low-level features and high-level concepts (Zhu & Liu, 2009b).

Recently, in order to bridge the gap, visual attention model-based summarization schemes and EEG-based video summarization methods were proposed (Mehmood, Sajjad, Rho, & Baik, 2016; Ejaz, Mehmood, & Baik, 2014; Kim & Kim, 2016). In the following sections, we explain an overview of previous studies on video summarization methods, and also detail two studies: an algorithm for key-frame extraction, and a video summarization method using EEG/ERP techniques.

7.2 Related Studies

Yang and Marchionini (2004) found that users liked to use visual surrogates for relevance judgments, especially those surrogates that contained motion, although topicality belonging to textual relevance criteria was still considered the most important criteria for video relevance judgments. Iyer and Lewis (2007) investigated the ability of video storyboards to summarize and communicate the themes of arts−related videos. Their research results presented the importance of storyboards as surrogates for videos: 75% of the responses indicated that the participants considered storyboards useful in deciding whether to view a full−length video. However, they questioned the linear sequence and narrative structure of a storyboard that had no thematic links between key−frames. Thus, they proposed a model to improve the storyboard's ability to communicate the essential message of videos.

Kim (2007) examined whether visual storyboard surrogates are sufficiently useful to be utilized as sources for indexing and searching a video. The study showed that the effectiveness of video storyboards as sources for both video indexing and searching varies according to the type of video: If a video conveys its meanings primarily through images, then the effectiveness will be high; however, if a video conveys its meanings primarily through narration, then the effectiveness will be low.

Song and Marchionini (2007) compared the effectiveness of three different surrogates: visual alone (storyboard), audio alone (spoken description), and a combination of video and audio (a storyboard augmented with spoken description). The study showed that combined surrogates are more effective and, hence, strongly preferred. They also demonstrated that the use of only oral descriptions led to better comprehension of the video segments than only visual storyboards; however, people preferred to have visual surrogates and use them to confirm interpretations and add context. Marchionini, Song, and Farrell (2009) examined the effectiveness of audio surrogates alone and in combination with one kind of visual surrogate, fast forwards. Their research results showed that manually generated spoken descriptions are better than both manually generated spoken keywords and fast forwards for video gisting. That is, high quality spoken summaries were found to be very effective for video gisting, and the visual elements added subjective value to a user's experience.

7.3 Algorithm for Key-frame Extraction

We proposed the extraction of key frames using a two-step

process (Kim & Kim, 2009; Kim & Kim, 2010). In the first step, we begin by pre—sampling frames from each video. In the second step, we refine our selection of key frames, manually, on the basis of scene—and video—level attributes. At the video level, we select key frames having overlay text or narrators and then group the selected frames into scenes, whereas at the scene level, we select frames that include objects, events/actions, people, or scenery.

We select six types of key frames (overlay text, narration, object, event/action, person, and scenery), because analysis of the results of the previous studies and our own preliminary study show that people rely heavily on identifying not only visual information such as objects or events but also verbal information such as overlay text or letters on objects, when making sense of images (or videos).

We select key frames at the scene level instead of at the shot level because we assume that a scene composed of several shots, creatively combined together, enables viewers to extract more meaning from the content of a video (Smeaton, 2007). Many previous studies have investigated methods for scene detection, though the best approach to scene formulation, with respect to ensuring human understanding, remains an open question (Zhu & Liu, 2009a). Key frame selection at the scene level requires an analysis of the story structure of a video and subsequent segmentation into story elements (Smeaton, 2007). Toward this end, we suggest criteria for use in the detection of scene boundaries, taking into account aspects of human understanding and cognition, as well as the results of the abovementioned studies.

7.3.1 Two-step Approach to Key-frame Extraction

The following section describes the proposed two-step approach to key-frame extraction.

Step one: We pre-sample frames from each video to establish a subset of candidate frames by employing a mechanical method, in which key frames for each video are extracted every 3 s. This pre-sampling allows us to reduce the number of initial, candidate key frames to a manageable volume. Prior work has shown that the quality of summarization is not affected by pre-sampling carried out in this manner (Mundur et al., 2006).

Step two: We manually select key frames from the remaining candidates, on the basis of two different levels of attributes: video- and scene-levels. While selection at the video level is based on the presence of overlay text or narrators, at the scene level, frames that include objects, events, persons, or scenery are selected.

The abovementioned two-step process is performed in the following six phases:

1) For each video, extract a set of candidate key frames to work with, by mechanically stripping out one frame every 3 s using a

KMPlayer (Konqueror Media Player). The resulting set of frames will include one or two frame(s) per shot, as each shot typically lasts 2~5 s.

2) From the remaining set, using video–level criteria, select frames having overlay text, schematics, or symbols (T–frames). If redundant T–frames appear, then select only those T–frames that appear in the opening or ending sequences.

3) Again, using video–level criteria, select one frame that includes one or more narrators (N–frames) from the set of candidate key frames. If redundant N–frames appear, then select the one that appears first.

4) Manually group the candidate key frames into scenes. By adopting the notion of an establishing shot, which is defined as a shot that establishes a scene's setting and/or participants, we set two criteria for the detection of scene boundaries. First, we identify scene boundaries on the basis of scenery changes, where some variation in the spatial background of a scene is observable. Second, we identify scene boundaries on the basis of changes in scene actors, where different individual(s) are observed in the scene from one point to the next.

5) Select object frames (O–frames), action/event frames (E–frames), person frames (P–frames), and scenery frames (S–frames) that indicate the central theme of a video. Here, the

scenery frames present the geographical settings or time periods that are related to key events or actions in the video. If several E−frames are used to create an action or event, then select multiple E−frames. Next, if multiple redundant P−frames are present, select the P−frame that includes the greatest number of key individuals. We do not specify an order for the selection of frames of each type, because it is dependent on video genre; some genres will require an emphasis on a certain person or event.

6) Select a pair of O−frames, E−frames, P−frame, S−frames, or any combination of the above, if making a connection between them can be effectively used to determine the subject of a video.

7.3.2 Example of Storyboard Construction

On the basis of the proposed criteria for key−frame extraction and display, 12 sample videos were analyzed. Two researchers independently selected a group of key frames from each video by employing the proposed algorithm. The average rate of corres− pondence between the two groups of key frames identified by the researchers was 0.71. In each case of disagreement between the frame selections of the two researchers, effort was made to reach a consensus on the final result.

By way of an example, we explain below how the storyboard of Video 6 ("Computer Rage," having a duration of 2 min 56 s) was constructed. Following pre-sampling, the 4,262 frames included in Video 6 were reduced to 56 candidate key frames. From these 56 candidate key frames, a T–frame (v01) and an N–frame (v30) were selected. After grouping the key frames into 8 scenes, three O–frames (v23, V29, and v43) were selected from the 4th and 7th scenes; four E–frames (v05, v18, v24, and v48) from the 1st, 2nd, 4th, and 7th scenes; and three P–frames (v11, v19, and v26) from the 2nd, 3rd,

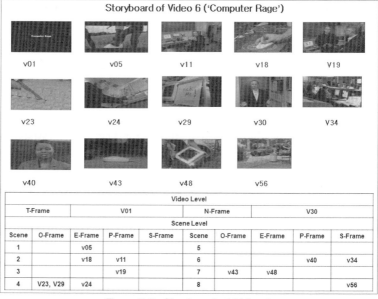

Storyboard of Video 6 ('Computer Rage')				
v01	v05	v11	v18	V19
v23	v24	v29	v30	V34
v40	v43	v48	v56	

Video Level									
T-Frame		V01			N-Frame		V30		
Scene Level									
Scene	O-Frame	E-Frame	P-Frame	S-Frame	Scene	O-Frame	E-Frame	P-Frame	S-Frame
1		v05			5				
2		v18	v11		6			v40	v34
3			v19		7	v43	v48		
4	V23, V29	v24			8				v56

Figure 7-2. Storyboard of Video 6

and 4th scenes. Additionally, two S-frames (v34 and v56) were selected from the 6th and 7th scenes. As a result, we constructed a storyboard (Figure 7-2) for Video 6 that consists of 14 key frames.

7.4 Video Summarization Using EEG/ERP Techniques

7.4.1 Overview

As mentioned above, many keyframe extraction methods were proposed that use the shot-level or scene-level boundary detection techniques (Smeaton, Over, & Doherty, 2010; Chen, Delannay, & Vleeschouwer, 2011; Mishra, Raman, Singhai, & Sharma, 2015), and the content features of a video (Tavassolipour, Karimian, & Kasaei, 2014).

Despite rapid improvement in boundary detection and content feature extraction algorithms, the results of these studies still remain unsatisfactory owing to a semantic gap (Hu, Xie, Li, Zeng, & Maybank, 2011), which refers to the gap between the actual semantic topic that a viewer infers from watching a video and the topic inferred from video summarization, which is constructed using the extraction algorithms.

Recently, visual attention model-based summarization schemes were proposed that extract frames as keyframes if they are visually important for humans as determined by visual attention models (Mehmood et al, 2016; Ejaz et al., 2014). However, it seems that the real issue underlying the semantic gap does not refer to the visual attention, but rather to how viewers respond to incoming multimedia content, pondering upon the semantic meanings of the stimuli and integrating those meanings into a coherent mental representation.

Human attention to incoming visual stimuli is assumed to be automatic and its perception to be near–automatic in encoding formal features of the stimuli. Thus, the semantic gap does not seem to be related with this automatic attention or near–automatic perception phases, but to be related with meanings of those stimuli and with integrating those meanings into a coherent mental representation.

In order to bridge the semantic gap, we developed a method that extracts topic–relevant shots using EEG data obtained during a video–watching session. For this end, we reviewed studies on event–related potential (ERP) components that are used to measure users' cognitive responses evoked by words, static images, or silent video clips. Based on this related studies, we proposed and tested two hypotheses. We then evaluated our proposed video summarization method and described the evaluation results in Chapter 9.3 entitled "The Evaluation of EEG–based Key Shot Extraction Algorithm."

7.4.2 Theoretical Background

1) N400 and P600 Effects

According to the contextual updating theory in linguistic literature, once some kind of topical expectation is obtained by the preceding words in a sentence and if a newly placed word makes a mis—match with this topic expectation, then there comes a very specific responses to this mis—match: N400 effects (Friedman & Johnson, 2000). Let us look at the relation between the phrase 'a type of fish' and the word 'dolphin'. Subjects are looking forward to a type of fish, say 'salmon', and when they hear something other than a fish, then their expectation makes a mis—match with this latest word, 'dolphin'. Let us look at another case for P600. P600 effect is considered as brain wave responses which reflect the cognitive effort to process the stimulus information that is difficult to handle with.

2) Semantic Illusion: P600 without N400

The fact that P600 effect exists without N400 effect was often found in linguistic experiments. Researchers understand it as a result of semantic illusion in which syntactic error or others are not perceived as such an error to deliver semantic information (Kim & Osterhout, 2005). Semantic illusion was also found in the context wider than just syntactic errors. Let us take an example. In a lab experiment, when asked the number of animals in the Moses' Ark, subjects tended to answer to this question by using prior knowledge of 'Noah's Ark' (Brouwer, Fitz, & Hoeks, 2012). It is referred as 'Moses illusion'. Researchers explain the semantic illusion by the notion of semantic attraction (Kim & Osterhout, 2005). Semantic attraction theory explains that more attractive semantic meaning can command audience's attention than less attractive syntactic errors, allowing semantic integration to occur while disregarding syntactic errors. It means that if semantic communication matters, syntactic error doesn't matter at all.

7.4.3 Research Hypotheses: Semantic Mismatch and Integration

We treated visual shots in a video as consecutively presented stimuli introduced to viewers over shot—boundaries or cuts between visual shots. Thus, in order to select topic—relavant shots, we decided to examine how viewers respond to topic—relevant or topic—irrelevant shots while watching a video. We focused on the integration phase and excluded the early attention and perception phases from our analysis. This is because the two early phases seem to be automatic or near—automatic responses to stimuli so that we do not expect significant differences between two different conditions in the two phases (Allegretti et al., 2015). We assumed that the integration phase consists of two steps: a semantic mismatching process for topic-irrelevant shots, and a context updating process for topic-relevant shots. We constructed a semantic mismatch and integration model (Figure 7-3), which shows that each of two steps is linked to a specific ERP component.

1) Hypothesis 1

The N400 component is specific to the semantic mismatch for topic-irrelevant shots (Hamm et al., 2002), as in linguistics experiments. For example, if incoming topic-irrelevant shots do not match prior topical expectation and thus do not need further semantic integration, then low negative ERP signals are elicited at around 400 ms. We hypothesize that an N400 effect occurs for topic-irrelevant shots and elicits a more negative ERP signal in response to topic-irrelevant shots compared to topic-relevant shots.

2) Hypothesis 2

The P600 component has been suggested to reflect topic shift (Hung & Schumacher, 2012), maintaining and updating discourse structure (Schumacher & Hung, 2012), and discourse-internal reorganization and integration (Wang & Schumacher, 2013).

Considering the findings from these studies, we can expect that a P600 effect occurs when topic—relevant shots correspond to prior topical expectations and thus need further semantic integration. In other words, the P600 component is used to describe the context—updating step for topic—relevant shots wherein incoming concepts are integrated with existing concepts, and with prior knowledge from the long—term memory. It is thus hypothesized that a P600 effect appears for topic—relevant shots and produces a more positive ERP signal for topic—relevant shots than for topic—irrelevant shots.

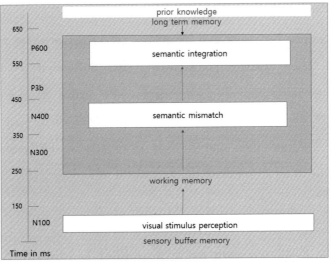

Figure 7-3. Semantic mismatch and integration model

7.4.4 Experimental Methodology

1) Tasks and Experimental Materials

We conducted two experiments using a within–subject design wherein the independent variable was relevance. This research was approved by the Institutional Review Board of Myongji University. Written informed consent was obtained from each participant prior to his or her involvement in the experimental sessions.

We decided to use videos in the same genre (documentary videos) as experimental materials because the characteristics of videos seem to vary according to genre. We selected three videos from YouTube and Korea Heritage (http://www.k-heritage.tv/). The experimental materials used for the experiment are shown in Table 7–1. In our experiment, we also utilized twenty–seven shots extracted from the three videos as stimuli. We evaluated all shots belonging to each video and selected the three most relevant, three partially relevant, and three irrelevant shots from each video based on the degree of topic relevance. This resulted in the selection of nine shots per video.

Table 7-1. Experimental materials

No.	Title	Number of presented shots
1	Discovering glasses	9
2	Uhm's bicycle	9
3	Two pocket watches	9

2) Participants

We recruited 44 participants (35 undergraduate and 9 graduate students) from Myongji University by email and phone. The participants were compensated with money. The participants were limited to right−handed male students in the age range of 21~27 years to minimize differences between individual EEGs. We devided the participants into two gropus: 21 participants for the first experiment and 23 participants for the second experiment

3) Procedure

Figure 7−4 shows how the videos and their shots were presented to the participants. The participants were provided with instructions for the experiment. We asked the participants to focus on the topic of the video and to memorize shots reflective of the central theme of the video while watching the video.

In the first experiment using video watching session 1, three videos and twenty−seven shots extracted from them (nine shots per video) were used as stimuli. For example, one video (e.g., Video 2) randomly chosen from the three videos was shown to the participants. Before the presentation of the video, a 3−s black screen with a fixation cross was presented. This was followed by a 500−ms black screen. After watching the video, each of the nine shots in the video was sequentially presented for 2 s to the participants. Before the presentation of each shot, a 2−s black screen with a fixation cross was presented. This was followed by a 500−ms black screen.

In the second experiment using video watching session 2, three videos and their twenty−seven shots were also used as stimuli. The difference btween these two experiments is that in the second experiment, a 200−ms black screen was used in between shots during the video watching session in order to obtain pre−stimulus data (Figure 7−4), whereas in the first experiment, the 200 ms prior to each shot boundary was used as a pre−stimulus period instead of using a 200−ms black screen.

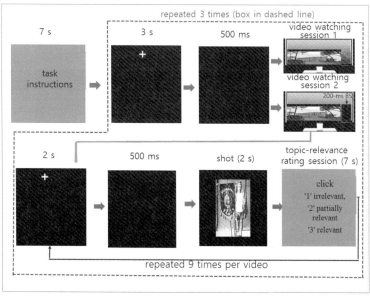

Figure 7-4. The order and timing of the experiment

We thus separated the topic—relevance rating session from the video—watching session. This is because if the topic—relevance rating was performed during the video—watching session, the button—pressing behavior may have evoked response—related ERP components (Luck, 2014). This would have contaminated ERP responses of interest to our research during the natural watching session.

Figure 7-5. The nine shots of Video 2 (Uhm's bicycle).

During the topic—relevance rating session, the participants were asked to rate each of the nine shots belonged to the video by pressing the "1" key on the computer keyboard if the presented shot was irrelevant to the topic of the video, the "2" key on the computer keyboard if it was partially relevant, or the "3" key on the computer keyboard if it was relevant. The participants were required to respond within 7 s. The above steps were repeated for the other two videos. We will describe more about the topic-relevance rating and video watching sessions in the following section.

4) EEG Recording and Data Analysis

We used a SynAmps amplifier linked to a Quik-Cap with 32 channels (Compumedics Neuroscan; Victoria, Australia) for EEG data acquisition. Data were digitally sampled for analysis at a 1,000-Hz sampling rate. EEG data were analyzed using a 7.09 CURRY program (Compumedics Neuroscan). The mean values from each participant's EEG data during the 200 ms before stimulus onset were adjusted to zero as a baseline correction for the participant's background brainwave. ERP brainwave data were collected during the 1,000 ms after stimulus onset. Thus, EEG data were analyzed mainly over a time period of 1,200 ms (-200 ms to 1,000 ms). We will describe below how to analyze the EEG signals acquired from the topic-relevance rating session and two video-watching sessions.

(i) Topic-relevance Rating Session

The EEG signals acquired from the topic-relevance rating session were preprocessed. The preprocessing steps consisted of setting a band-pass filter from 0.1 to 35 Hz, which was applied to remove power−line noise and extracting epochs from 200 ms before stimulus presentation to 1,000 ms after stimulus presentation. The EEG data for 27 shots (per participant) obtained from both experiments during the relevance−rating session were first collected. Among the 1,188 epochs (27 epochs per participant) obtained from the 44 participants, 386 epochs were found to be topic−relevant, 417 epochs were partially relevant, and 385 epochs were topic−irrelevant.

(ii) Video Watching Session 1

To obtain the EEG data for video shots observed while watching a video wherein a black screen was not used between shots, fifteen shots out of a total of 27 shots were selected. The shots that appeared in the first part of a video were excluded because we hypothesized that the participants may not have recognized the topic of the video at the beginning. From the remaining shots, we selected 15 shots with high agreement rates in the topic−relevance assessments by the 21 participants. These 15 shots were used to identify corresponding shots in the EEG video data file.

For example, we selected five shots (specifically, the second, fourth, fifth, sixth, and seventh shots) from Video 2 for the analysis of the video−watching session (Figure 7−5). The EEG signals acquired from the video−watching session were analyzed using relevance ratings obtained from the relevance−rating session.

Therefore, if a given shot was rated as topic—relevant in the relevance—rating session, then we regarded the video shot from which the given shot was extracted as topic—relevant. Using the same method as that used for the relevance—rating session, epochs were extracted from 200 ms before to 1,000 ms after shot presentation.

We obtained 315 epochs (15 epochs per participant) from the 21 participants, of which 101 epochs were found to be topic—relevant, 111 epochs were partially relevant, and 103 epochs were topic—irrelevant. We finally used only 204 topic—relevant and topic—irrelevant shots. Partially relevant shots were excluded due to their vague characteristics.

We conducted a discriminant analysis to examine how the relevance ratings match with real—time EEG signal waves of the 204 epochs obtained while watching a video; we found that the EEG wave patterns of 27 epochs out of 204 epochs did not match with the corresponding relevance ratings assigned to them later by participants.

There can be several causes for this discrepancy, in addition to the inaccuracies caused by small training samples used in the discriminant analysis. For example, participants often do not know exactly what a video is about until they watch the video; thus, they may ultimately rate a given shot (epoch) as irrelevant in the topic—relevance rating session conducted after watching the video, even though they thought the same shot was relevant while they were watching the video. In order to minimize the discrepancy, only 177 epochs (27 unmatched epochs excluded) were used in classifying video shots as relevant or irrelevant (refer Chapter 9.3.1 entitled "Discriminant Analysis: Variable Selection and Evaluation"). However, we used 204 epochs to verify two hypotheses described below.

Unlike the relevance−rating session, these shots were presented one after another without a black screen in between shots. This makes it difficult to define a baseline for the ERP measurements while the participant watching a video. In the first experiment using the video watching session 1, we used the last 200−ms time period prior to each shot boundary as the pre−stimulus time period, instead of using a black screen in between the shots. In the second experiment using the video watching session 2, a 200−ms black screen was used in between video shots, in order to obtain more accurate pre−stimulus data; that is, before the presentation of each shot of a video, a 200−ms black screen was presented. We will detail below the video watching session 2.

(iii) Video Watching Session 2

To obtain the EEG data for video shots observed while watching a video wherein a black screen was used between shots, fifteen shots out of a total of 27 shots were also selected. Using the same method as that used for the video watching session 1, epochs were extracted from 200 ms before to 1,000 ms after shot presentation. We obtained 345 epochs (15 epochs per participant) from the 23 participants, of which 108 epochs were found to be topic-relevant, 127 epochs were partially relevant, and 110 epochs were topic-irrelevant.

7.4.5 Results

1) Signal-to-Noise Ratio

In a standard ERP experiment, an experimental stimulus is presented after a brief presentation of a black screen. This pre-stimulus period usually occurs 200 ms before the stimulus onset. The background EEG is the baseline brain activation and is compared with a signal EEG to calculate the SNR, which is defined as the ratio of the signal EEG to the background EEG. The background EEG is operationalized as an EEG response with no external stimulus, whereas the signal EEG is an EEG response after the onset of a stimulus. However, when a video reveals itself one shot after another, as in natural video viewing, it raises an issue for the operational definition of the background EEG because there is some residue of the previous shot before the onset of the next shot.

To investigate how the insertion of a black screen affects the noise level and SNR of ERP signals obtained during video viewing, we compared the results of the topic−relevance rating session and video−watching session 2, wherein a black screen was used before the shot onset, to those of the video−watching session 1, wherein the 200 ms prior to each shot boundary was used as a pre−stimulus period. The shapes of the plots obtained from three sessions are similar (Figs. 7−6, 7−7, and 7−8). In the relevance−rating session, the noise level was 0.173 and the maximum SNR was 27.3, whereas in the the video−watching session 2, the noise level was 0.251 and the maximum SNR was 16.0. Meanwhile, in the video−watching session 1, the noise was 0.194 and the maximum SNR was 10.3. These SNR values were larger than the reasonable level of 10.0 (Luck, 2004). We conclude that the 200 ms prior to each shot boundary can be used as a pre−stimulus epoch.

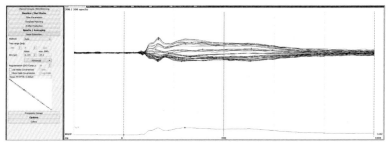

Figure 7-6. Butterfly plot of ERP responses for topic-relevance rating session. Upper graph (blue) represents overlapping ERP responses for 30 channels, and lower graph (red) represents MGFT (Grand Mean)(Noise=0.173, Max SNR=27.3)

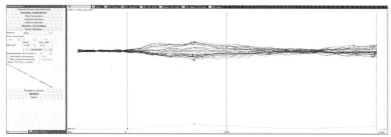

Figure 7-7. Butterfly plot of ERP responses for video watching session 1 (without blacks screens) (Noise=0.194, Max SNR=10.3).

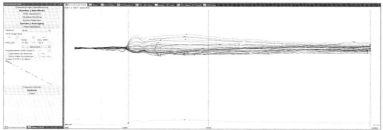

Figure 7-8. Butterfly plot of ERP responses for video watching session 2 (with blacks screens) (Noise=0.251, Max SNR=16.0).

2) ERP Results

The grand-average ERP waveforms at electrodes Cz and Fz elicited by the irrelevant shots (dashed line) have negative peaks around 500 ms (N400)(Figure 7-9). However, ERP waveforms elicited by the relevant shots (solid lines) have positive peaks around 600 ms (P600). We noticed that the overall time course of the N400 negativity in response to the irrelevant shots is more prolonged than that typical of N400 evoked by abruptly presented written words. This is consistent with the observations of Sitnikova et al. (2008), who reported that such results are not surprising because incongruous information unfolds over time in dynamic video scenes.

We performed a repeated−measures t−test with a within−subject factor of topic−relevance (1: topic−irrelevant, 3: topic−relevant) to verify the two hypotheses. We used only topic−relevant and topic−irrelevant shots.

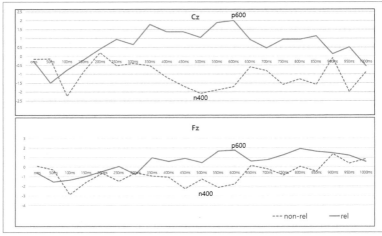

Figure 7-9. Grand-average ERP waveforms elicited at electrodes Fz and Cz associated with irrelevant shots (dashed) and relevant shots (solid) (204 epochs obtained from the video watching session 1 were analyzed for the grand-average ERP waveforms).

(i) Hypothesis 1

We verified the first hypothesis that an N400 effect appears for topic−irrelevant shots. More negative ERP signals were elicited in response to the topic−irrelevant shots than the topic−relevant shots. The t−test results indicate the presence of a clear N400 and a significant difference in N400 activation between the two cases (Table 7−2).

Table 7-2. Means of minimums at 350~600ms (N400) between topic-relevant and topic-irrelevant video shots (unit: μV)

CH	Mean (SD)		t (p)	CH	Mean (SD)		t (p)
	Irrelevant	Relevant			Irrelevant	Relevant	
F3	−8.51 (6.14)	−3.77 (4.25)	10.16 (.005)[**]	FP1	−9.35 (7.36)	−2.98 (5.77)	13.04(.002)[**]
Fz	−8.36 (5.39)	−3.31 (4.10)	11.34 (.003)[**]	FP2	−9.36 (7.35)	−3.02 (4.11)	14.25(.001)[***]
C3	−7.00 (5.46)	−3.03 (4.35)	8.79 (.008)[**]	FC3	−7.67 (5.76)	−3.57 (4.31)	8.53(.008)[**]
Cz	−7.23 (4.84)	−2.52 (3.92)	12.89 (.002)[**]	CP3	−5.75 (5.24)	−2.30 (3.95)	9.00(.007)[**]

[*] $p < 0.05$, [**] $p < 0.01$, [***] $p < 0.0017$ (after Bonferroni correction), SD: standard deviation

At the statistical significance level of 0.01, the more negative potentials for topic−irrelevant shots when compared to topic−relevant shots were evoked at the midline frontal and left frontal lobes (Fz and F3), the midline central and left central lobes (Cz and C3), the left pre−frontal and right pre−frontal regions (FP1 and FP2), the left frontal−central region (FC3), and the left central−parietal lobe (CP3). The difference in responses at the FP2 remained significant after the Bonferroni correction $(p < 0.0017)$.

(ii) Hypothesis 2

We verified the second hypothesis that a P600 effect appears for topic−relevant shots. More positive ERP signals are elicited in response to topic−relevant shots than topic−irrelevant shots. The t−test results indicate a significant difference in P600 activation between the two cases (Table 7−3).

Table 7-3. Means of maximums at 400~600ms (P600) between topic-relevant and topic-irrelevant video shots (unit: μV)

CH	Mean (SD)		t (p)	CH	Mean (SD)		t (p)
	Irrelevant	Relevant			Irrelevant	Relevant	
F3	1.19 (3.30)	5.00 (3.86)	11.78 (.003)"	C4	1.93 (2.84)	5.29 (2.91)	13.70 (.001)'''
Fz	0.91 (3.25)	5.16 (3.56)	17.99 (.000)'''	FP1	0.83 (4.56)	5.92 (6.38)	9.72 (.005)"
F4	0.27 (4.07)	4.33 (3.83)	12.72 (.002)"	FC3	1.186 (3.68)	4.48 (3.41)	8.62 (.008)"
C3	1.62 (3.55)	4.96 (2.67)	13.41 (.002)"	FCz	0.38 (3.49)	4.94 (3.36)	17.05 (.001)'''
Cz	0.80 (2.99)	5.66 (2.71)	29.24 (.000)'''	FC4	0.71 (3.28)	4.45 (2.86)	16.40 (.001)'''

'$p < 0.05$, "$p < 0.01$, "'$p < 0.0017$ (after Bonferroni correction), SD: standard deviation

At the statistical significance level of 0.01, the more positive potentials for topic−relevant shots than topic−irrelevant shots were evoked at the left frontal and right frontal lobes (F3 and F4), the midline frontal lobe (Fz), the left central and right central lobes (C3 and C4), the midline central lobe (Cz), the left pre−frontal region (FP1), the left frontal−central and right frontal−central regions

(FC3 and FC4), and the midline frontal–central region (FCz). Differences in responses at the Fz, Cz, C4, FCz, and FC4 remained significant after the Bonferroni correction ($p < 0.0017$).

References

Allegretti, M., Moshfeghi, Y., Hadjigeorgieva, M., Pollick, F. E., Jose, J. M., & Pasi, G. (2015). When relevance judgement is happening?: An EEG–based study. In Proceedings of the 38th International ACM SIGIR Conference on Research and Development in Information Retrieval (pp. 719~722). New York: ACM.

Anthimopoulos, M., et al. (2007). Detecting text in video frames. In Proceedings of the Fourth Conference on IASTED International Conference: Signal Processing, Pattern Recognition, and Applications, 39~44, Innsbruck, Austria.

Ariki, Y., Kumano, M., & Tsukada, K. (2003). Highlight scene extraction in real time from baseball live video. Proceedings of the Fifth ACM SIGMM International Workshop on Multimedia Information Retrieval (pp. 209~214). New York: ACM Press.

Baddeley, A. (2007). Working memory, thought and action. Oxford: Oxford University Press.

Beaudoin, J. (2007). Flickr image tagging: Patterns made visible. Bulletin of the American Society for Information Science and Technology, 34(1), 26~29.

Brouwer, H., Fitz, H., & Hoeks, J. (2012). Getting real about semantic illusions: Rethinking the functional role of the P600 in language comprehension. Brain Research, 1446, 127~143.

Chen, F., Delannay, D., & De Vleeschouwer, C. (2011). An autonomous framework to produce and distribute personalized team–sport video summaries: A basketball case study. IEEE Transactions on Multimedia, 13(6), 1381~1394.

Corbetta, M., & Shulman, G. (2002). Control of goal–directed and stimulus–driven attention in the brain. Nature Reviews Neuroscience, 3(3), 201~215.

Danisman, T., & Alpkocak, A. (2006). Dokuz Eylül University video shot boundary detection at TRECVID 2006. Proceedings of TRECVID 2006.

Detyniecki, M., & Marsala, C. (2007). Video rushes summarization by adaptive acceleration and stacking of shots. In Proceedings of the International Workshop on TRECVID Video Summarization, 65~69, Augsburg, Germany.

Ding, W., et al. (1999). Multimodal surrogates for video browsing. Proceedings of the Fourth ACM Conference on Digital Libraries (pp. 85~93). Berkeley CA, USA.

Dumont, E., & Merialdo, B. (2007). Split–screen dynamically accelerated video summaries. TVS 2007 –TRECVID BBC Rushes Summarization Workshop, ACM Multimedia.

Eakins, J. P., & Graham, M. E. (1999). Content–based image retrieval. JISC Technology Applications Programme Report, 39.

Ejaz, N., Mehmood, I., & Baik, S. W. (2014). Feature aggregation based visual attention model for video summarization. Computers & Electrical Engineering, 40(3), 993~1005.

Eugster, M. J., Ruotsalo, T., Spapé, M. M., Kosunen, I., Barral, O., Ravaja, N., ... & Kaski, S. (2014). Predicting term–relevance from brain signals. In SIGIR 14 conference committee (Eds.), Proceedings of the 37th International ACM SIGIR Conference on Research & Development in Information Retrieval (pp. 425~434). New York: ACM Press.

Friedman, D., & Johnson, R. (2000). Event-related potential (ERP) studies of memory encoding and retrieval: A selective review. Microscopy Research and Technique, 51(1), 6~28.

Geisler, G., & Burns, S. (2007). Tagging video: Conventions and strategies of the YouTube community. Proceedings of the Joint Conference on Digital Libraries (JCDL 2007), p. 480 [poster].

Gola, M., Kamiński, J., Brzezicka, A., & Wróbel, A. (2012). Beta band oscillations as a correlate of alertness–changes in aging. International Journal of Psychophysiology, 85(1), 62~67.

Goldenholz, D. M., Ahlfors, S. P., Hämäläinen, M. S., Sharon, D., Ishitobi, M., Vaina, L. M., & Stufflebeam, S. M. (2009). Mapping the signal-to-noise-ratios of cortical sources in magnetoencephalography and electroencephalography. Human Brain Mapping, 30(4), 1077~1086.

Golder, S., & Huberman, B. A. (2005). The structure of collaborative tagging systems. HP Labs Technical Report. Available from http://www.hpl.hp.com/research/idl/papers/tags/

Greisdorf, H., & O'Connor, B. (2002). Modelling what users see when they look at images: A cognitive viewpoint. Journal of Documentation, 58(1), 6~29.

Hamm, J. P., Johnson, B. W., & Kirk, I. J. (2002). Comparison of the N300 and N400 ERPs to picture stimuli in congruent and incongruent contexts. Clinical Neurophysiology, 113(8), 1339~1350.

Holcomb, P. J., & McPherson, W. B. (1994). Event-related brain potentials reflect semantic priming in an object decision task. Brain and Cognition, 24(2), 259~276.

Hu, W., Xie, N., Li, L., Zeng, X., & Maybank, S. (2011). A survey on visual content-based video indexing and retrieval. IEEE Transactions on Systems, Man, and Cybernetics, Part C (Applications and Reviews), 41(6), 797~819.

Hughes, A., et al. (2003). Text or pictures? An eyetracking study of how people view digital video surrogates. In Proc. Conf. Image and Video Retrieval (CIVR) (Urbana-Champaign, IL, 2003), 271~280.

Hung, Y.-C., & Schumacher, P. B. (2012). Topicality matters: Position-specific demands on Chinese discourse processing. Neuroscience Letters, 511(2), 59~64.

Ingwersen, P. (2002). Cognitive perspectives of document representation. CoLIS 4: 4th International Conference on Conceptions of Library and Information Science (pp. 285~300). Greenwood Village: Libraries Unlimited.

Iyer, H., & Lewis, C. D. (2007). Prioritization strategies for video storyboard keyframes. Journal of American Society for Information Science and

Technology, 58(5), 629~644.

Kaan, E., & Swaab, T. (2003). Repair, revision, and complexity in syntactic analysis: An electrophysiological differentiation. Journal of Cognitive Neuroscience, 15(1), 98~110.

Kim, H. (2007). An experimental study on the effectiveness of storyboard surrogates in the meanings extraction of digital videos. Journal of Korean Society for Information Management, 24(4), 53~72.

Kim, H., & Kim, Y. (2009). A two-step model for video key-frame determination. Proceedings of the Association for Information Science and Technology, 46(1), 1~16.

Kim, H., & Kim, Y. (2010). Toward a conceptual framework of key-frame extraction and storyboard display for video summarization. Journal of the Association for Information Science and Technology, 61(5), 927~939.

Kim, Y., & Kim, H. (2016). Semantic gap of video summarization and semantic illusion for visual information integration: Testing N400 and P600 effect hypotheses of topic relevance using ERP responses to real-time video watching. In Proceedings of the Annual Conference on Management and Social Sciences (pp. 209~218). Hawaii, USA.

Kim, A., & Osterhout, L. (2005). The independence of combinatory semantic processing: Evidence from event-related potentials. Journal of Memory and Language, 52(2), 205~225.

Koelstra, S., Mühl, C., & Patras, I. (2009). EEG analysis for implicit tagging of video data. Proceeding of the 3rd International Conference on Affective Computing and Intelligent Interaction and Workshops, ACII

2009 (pp. 27~32). IEEE Computer Society Press.

Laine—Hermandez, M., & Westman, S. (2008). Multifaceted image similarity criteria as revealed by sorting tasks. Proceedings of the ASIST Annual Meeting, 45, Columbus, Ohio.

Liu, Z., et al. (2007). AT&T research at TRECVID 2007. Proceedings of TRECVID 2007.

Luck, S. J. (2014). An introduction to the event—related potential technique. MIT press.

Marchionini, G., & Geisler, G. (2002). The Open Video Digital Library. D—Lib Magazine, 8(12).

Marchionini, G., Song, Y., & Farrell, R. (2009). Multimedia surrogates for video gisting: Toward combining spoken words and imagery. Information Processing and Management, 45(6), 615~630.

Marchionini, G., Wildemuth, B. M., & Geisler, G. (2006). The Open Video Digital Library: A Möbius strip of research and practice. Journal of American Society for Information Science and Technology, 57(12), 1629~1643.

Martí, M., Hinojosa, J. A., Casado, P., Muñoz, F., & Fernández—Frí, C. (2004). Electrophysiological evidence of an early effect of sentence context in reading. Biological Psychology, 65(3), 265~280.

Mayer, R. E. (2005). Cognitive theory of multimedia learning. In R.E. Mayer (Ed.), The Cambridge handbook of multimedia learning (pp. 134~146). New York: Cambridge University Press.

McPherson, W. B., & Holcomb, P. J. (1999). An electrophysiological investigation of semantic priming with pictures of real objects.

Psychophysiology, 36(1), 53~65.

Mehmood, I., Sajjad, M., Rho, S., & Baik, S. W. (2016). Divide−and−conquer based summarization framework for extracting affective video content. Neurocomputing, 174, 393~403.

Mishra, R., Raman, C. V., Singhai, S. K., & Sharma, M. (2015). Real time and non real time video shot boundary detection using dual tree complex wavelet transform. 2015 International Conference on Industrial Instrumentation and Control (ICIC, pp. 1495~1500). IEEE.

Moshfeghi, Y., Pinto, L. R., Pollick, F. E., & Jose, J. M. (2013). Understanding relevance: An fMRI study. In P. Serdyukov et al., eds. European Conference on Information Retrieval (pp. 14~25). Springer Berlin Heidelberg.

Mostafa, J., & Gwizdka, J. (2016). Deepening the role of the user: Neuro−physiological evidence as a basis for studying and improving search. In Proceedings of the 2016 ACM on Conference on Human Information Interaction and Retrieval (pp. 63~70). New York: ACM.

Mundur, P., Rao, Y., & Yesha, Y. (2006). Keyframe−based video summariz−ation using Delaunay clustering. International Journal on Digital Libraries, 6(2), 219~232.

Nakano, H., Rosario, M. A. M., Oshima−Takane, Y., Pierce, L., & Tate, S. G. (2014). Electrophysiological response to omitted stimulus in sentence processing. NeuroReport, 25(14), 1169~1174.

Panofsky, E. (1955). Meaning in the visual arts: Meaning in and on art history. Doubleday.

Paivio, A. (1986). Mental representations. New York: Oxford University Press.

Peters, I., & Stock, W. G. (2007). Folksonomy and information retrieval. Proceedings of the ASIST Annual Meeting, 44, Milwaukee, Wisconsin.

Polich, J. (2007). Updating P300: An integrative theory of P3a and P3b. Clinical Neurophysiology, 118(10), 2128~2148.

Quintarelli, E. (2005). Folksonomies: Power to the people. ISKO Italy UniMIB meeting, Milan, June 24, 2005.

Rasheed, Z., & Shah, M. (2005). Detection and representation of scenes in videos. IEEE Transactions on Multimedia, 7(6), 1097~1105.

Rorissa, A., & Iyer, H. (2008). Theories of cognition and image categorization: What category labels reveal about basic level theory. Journal of American Society for Information Science and Technology, 59(9), 1383~1392.

Ruchkin, D. S., Johnson, R., Mahaffey, D., & Sutton, S. (1988). Toward a functional categorization of slow waves. Psychophysiology, 25(3), 339~353.

Ruotsalo, T., Jacucci, G., Myllymäki, P., & Kaski, S. (2015). Interactive intent modeling: Information discovery beyond search. Communications of the ACM, 58(1), 86~92.

Sato, T., et al. (1998). Video OCR for digital news archive. In Proceedings Workshop on Content-based Access of Image and Video Databases, 52~60, Los Alamitos, USA.

Schumacher, P. B., & Hung, Y.-C. (2012). Positional influences on information packaging: Insights from topological fields in German.

Journal of Memory and Language, 67(2), 295~310.

Sen, S., et al. (2006). Tagging, communities, vocabulary, evolution. Proceedings of the 2006 20th Anniversary Conference on Computer Supported Cooperative Work, 181~190, Banff, Alberta, Canada.

Shatford, S. (1986). Analyzing the subject of a picture: A theoretical approach. Cataloging & Classification Quarterly, 6(3), 39~62.

Sitnikova, T., Kuperberg, G., & Holcomb, P. J. (2003). Semantic integration in videos of real-world events: An electrophysiological investigation. Psychophysiology, 40(1), 160~164.

Sitnikova, T., Holcomb, P. J., Kiyonaga, K. A., & Kuperberg, G. R. (2008). Two neurocognitive mechanisms of semantic integration during the comprehension of visual real-world events. Journal of Cognitive Neuroscience, 20(11), 2037~2057.

Smeaton, A. F. (2007). Techniques used and open challenges to the analysis, indexing and retrieval of digital video. Information Systems, 32, 545~559.

Smeaton, A. F., & Browne, P. (2006). A usage study of retrieval modalities for video shot retrieval. Information Processing and Management, 42(5), 1330~1344.

Smeaton, A. F., et al. (2008). Content-based video retrieval: Three example systems from TRECVid. International Journal of Imaging Systems and Technology, 18(2/3), 195~201.

Smeaton, A. F., Over, P., & Doherty, A. R. (2010). Video shot boundary detection: Seven years of TRECVid activity. Computer Vision and Image Understanding, 114(4), 411~418.

Song, Y., & Marchionini, G. (2007). Effects of audio and visual surrogates for making sense of digital video. Proceedings of CHI 2007, 867~876, San Jose, CA.

Tavassolipour, M., Karimian, M., & Kasaei, S. (2014). Event detection and summarization in soccer videos using bayesian network and copula. IEEE Transactions on Circuits and Systems for Video Technology, 24(2), 291~304.

Thornhill, D. E., & Van Petten, C. (2012). Lexical versus conceptual anticipation during sentence processing: Frontal positivity and N400 ERP components. International Journal of Psychophysiology, 83(3), 382~392.

Togawa, H., & Okuda, M. (2005). Position-based keyframe selection for human motion animation. Proceedings of 11th International Conference on Parallel and Distributed Systems, 82~185, Fukuoka, Japan.

Uchida, S., Miyazaki, H., & Sakoe, H. (2008). Mosaicing-by-recognition for video-based text recognition. Pattern Recognition, 41, 1230~1240.

van Berkum, J. J., Brown, C. M., Zwitserlood, P., Kooijman, V., & Hagoort, P. (2005). Anticipating upcoming words in discourse: Evidence from ERPs and reading times. Journal of Experimental Psychology: Learning, Memory, and Cognition, 31(3), 443.

van Berkum, J. J., Hagoort, P., & Brown, C. M. (1999). Semantic integration in sentences and discourse: Evidence from the N400. Journal of Cognitive Neuroscience, 11(6), 657~671.

Wang, L., & Schumacher, P. B. (2013). New is not always costly: Evidence from online processing of topic and contrast in Japanese. Frontiers in Psychology, 4, 363.

West, W. C., & Holcomb, P. J. (2002). Event—related potentials during discourse—level semantic integration of complex pictures. Cognitive Brain Research, 13(3), 363~375.

Yang, M., & Marchionini, G. (2004). Exploring users' video relevance criteria—A pilot study. Proceedings of the Association for Information Science and Technology, 41(1), 229~238.

Yokoi, K., Nakai, H., & Sato, T. (2008). Toshiba at TRECVID 2008: Surveillance event detection task. Proceedings of TRECVID 2008.

Yoon, J. (2008). Searching for an image conveying connotative meanings: An exploratory cross—cultural study. Library and Information Science Research, 30, 312~318.

Zhu, X., Goldberg, A. B., Eldawy, M., Dyer, C. R., & Strock, B. (2007). A text—to—picture synthesis system for augmenting communication. In Association for the Advancement of Artificial Intelligence (Vol. 7, pp. 1590~1595).

Zhu, S., & Liu, Y. (2009a). Automatic scene detection for advanced story retrieval. Expert Systems with Applications, 36, 5976~5986.

Zhu, S., & Liu, Y. (2009b). Video scene segmentation and semantic representation using a novel scheme. Multimedia Tools and Applications, 42, 183~205.

Chapter 8

Social Information Retrieval

8. Social Information Retrieval

This chapter describes social information retrieval in terms of its definition and how it is connected to social data, and introduces case studies on social indexing and search of videos.

8.1 Three Categories of Social Information Retrieval

Web 2.0 has allowed a new freedom for a user in his or her relationship with the Web by facilitating his interactions with other users who have similar tastes or share similar resources. With the advent of the social Web, the word social information retrieval emerged. Social information retrieval can be defined as the process of leveraging social relationships and social content to perform an information retrieval task with the objective of improving the efficiency of information retrieval (Bouadjenek, Hacid, & Bouzeghoub, 2016). Social information retrieval is composed of three categories: social indexing, social search, and social recommendation (see Figure 8-1), which was reconstructed on the basis of the model proposed by Bouadjenek et al. (2016).

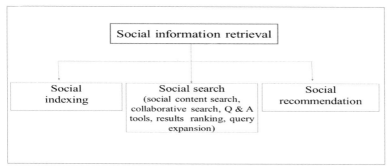

Figure 8-1. Three categories of social information retrieval

Let us take a look at three categories. Social indexing (tagging) is a distributed practice performed by a group of users when they assign uncontrolled keywords (tags) (Pan, He, Zhu, & Fu, 2016). Social search refers to finding information with the assistance of social resources, such as by asking friends, or unknown persons online for assistance (Morris, Teevan, & Panovich, 2010), and to use social information to improve the classic information retrieval process, e.g., document re-ranking and query reformulation. Social search also includes social content search and collaborative search. Social recommendation refers to a set of techniques that attempt to suggest items (e.g., books), social entities (e.g., persons), or topics of interest that are likely to be of interest to a user through the use of social information (Bouadjenek et al, 2016).

8.1.1 Social Indexing

Users have been made capable of describing documents, images, or videos with subject headings or "tags." This tagging reflects the users' conceptual model of information, and the tags present authentic representations of the language of authors and users (Quintarelli, 2005; Peters & Stock, 2007). Transforming the creation of explicit metadata for resources from a professional activity into a shared, communicative activity by users is an important development that should be explored and considered for future systems development (Mathes, 2004). Social indexing means the process of assigning social metadata to documents; here, social metadata refers to additional information about a resource resulting from user contributions and online activity—such as tagging, comments, reviews, ratings, recommendations—that helps people find, understand, or evaluate content (Smith—Yoshimura, 2011).

There are many studies regarding social indexing. Weller (2007) compared ontologies and folksonomies, and suggested that they are not to be seen as rivals but complementary to each other. Noruzi (2007) mentioned seven arguments for why a folksonomy—based system should use a thesaurus, emphasizing that it is impossible to maintain consistency over time or across folksonomy users without a thesaurus.

Peters and Stock (2007) explained Panofsky's theory by referring to an example of a photo found on Flickr and that Flickr image tags describe Panofsky's three levels (ofness, aboutness, and iconology) and aspects of "isness," referring to the work of Ingwersen (2002). Geisler and Burns (2007) classified tags assigned to YouTube videos into four categories: factual (93%), subjective (4%), personal (1%), and other (2%), on the basis of the categorizations developed by Sen et al. (2006) and Golder and Huberman (2005). The authors found that YouTube tags provide real added value, especially for user searches, because 66% of them do not appear in the other metadata (title or description).

Morrison (2008) compared the search performance of folksono-mies in information retrieval from social bookmarking Web sites with that of search engines and subject directories in terms of recall and precision rates. The results of this study showed that search engines had the highest precision and recall rates; however, folksonomies fared surprisingly well.

Melenhorst et al. (2008) explored the contribution of social tags, professional metadata, and automatically generated metadata for a retrieval process. One hundred ninety four participants were asked to tag a total of 115 videos, while another 140 participants were asked to search a video collection for answers to eight questions. The results demonstrated that social tags can result in an equally, if not more, effective search process than professionally generated metadata or automatically generated metadata.

Kim and Kim (2009) classified sampled Flickr tags according to their proposed tag category framework. The authors found that the most frequently used subcategories are those of location (12.93%) and object (12.45%), and that as many of the image tag terms are related to location, objects, and people, they would prove useful as index terms.

Yoo and Suh (2010) observed that the semantics of tags are not obvious because there is no hierarchy or relationship among tags. In order to minimize these problems, they suggested a user–categorized tag that freely defines the category of the tag by the user; based on these tags, a structured folksonomy was automatically created. Then, they developed prototype systems to investigate the utility of the structured folksonomy.

For the purpose of emotional tagging, Knautz, Siebenlist, and Stock (2010) designed a user interface that consisted of one picture and 18 slide controls on one web page; the 18 slide controls were split into two groups of 9 slide controls each. One group covered the emotions shown to a viewer when the picture was displayed. The other group contained the emotions felt by the viewer when looking at the picture. The range of every scroll bar contained values from 0 to 10, so the viewer could differentiate the intensity of every emotion.

Lee and Schleyer (2012) addressed the problem of evolutionary taxonomy construction in collaborative tagging systems and proposed an evolutionary taxonomy framework that consisted of two novel contributions in order to solve the problem. The authors conducted an extensive performance study using a large real-life web page tagging dataset (e.g., Del.ici.ous), verifying the effectiveness and efficiency of their proposed approach. Syn and Spring (2013) showed that social tags can be used as metadata and presented metrics to select the tags which have that potential.

Choi (2014) examined user-generated social tags in the context of subject indexing in order to determine how they could be used to organize information in a digital environment. The author employed the method of modified vector-based indexing consistency density with three different similarity measures: cosine similarity, dot product similarity, and Euclidean distance metric. The author found that social tags are more accurate descriptions of resources and a reflection of more current terminology than controlled vocabulary.

Golub, Lykke, and Tudhope (2014) explored the potential of applying the Dewey Decimal Classification as an established knowledge organization system for enhancing social tagging. The authors suggested that tagging enhanced with suggestions from DDC or another well-established knowledge organization system can make important improvements to the usual simple tagging, especially if the improvements suggested in their study are implemented.

Lee et al. (2014) analyzed the relativity of privacy preservation based on social tagging, and found the corresponding criteria of privacy relativity, such as publicity and privacy, interactivity, and independency. In fact, the tags actively annotated by users not only can reflect the personal preferences of relevant users but also can reveal the features of relevant information resources, whereas users can interact in the process of referring to each other and reannotating the same information resources. Thus, social tagging may have the characteristics of publicity and privacy, interactivity, and independency at the same time, and can be used to analyze the relativity of privacy preservation.

Recently, Cui, Shen, and Ma (2015) investigated the problem of tag relevance estimation from a new perspective of learning to rank and developed a novel approach to facilitate tag relevance estimation to directly optimize the ranking performance of tag-based image searches. The authors demonstrated the effectiveness of their approach in both the scenarios of image search and recommendation.

Pan et al. (2016) explored the usage patterns and regularities of co-employment of different popular tags and their relationships to the activeness of users and the interest level of resources in social tagging in order to examine the potential of social tagging in organizing, managing, sharing, and discovering large-scale online resources. The authors suggested that multiple individual tags and one or very few popular tags are generally employed together in one tagging action, and the usage patterns and regularities of tags with varying popularity are correlated to both user activity and resource interest.

8.1.2 Social Search

A social search is defined as the process of finding information using social resources and social relevance and includes social content search, collaborative search, Q & A tools, results ranking, and query reformation. Users can publish, provide, and spread information (e.g., commenting about an event) through social platforms, such as Twitter and Facebook. In such a context, a huge quantity of social information (social content, social relationship, social metadata) is created in social media. Hence, many users search social content to gather recent information about a particular person or event by using social content search engines, such as TwitterSearch and Facebook Search. Figure 8-2 shows the Facebook search result.

Most search engines are designed for a single user who searches alone; thus, users cannot benefit from the experience of each other for a given search task. Filho, Olson, and de Geus (2010) proposed a

Figure 8-2. Facebook search result

search interface that was designed to facilitate information seeking for inexperienced users by allowing more experienced users to collaborate together. SearchTeam is a concrete example of a collaborative search system currently available. Collaborative search in SearchTeam is conducted within a SearchSpace. A user begins his or her research by creating a SearchSpace on a topic of interest and then finds and saves only what he or she wants during searching and discards what he or she does not want or finds irrelevant.

Teevan, Ramage, and Morris (2011) explored the search behavior on the popular microblogging/social networking site Twitter. Using the analysis of large−scale query logs and supplemental qualitative data, the authors observed that people search Twitter to find temporally relevant information (e.g., breaking news) and information related to people (e.g., information about people of interest) and Twitter queries are shorter, more popular, and less likely to evolve as a part of a session than Web queries.

Wang, Luo, and Yu (2016) proposed a learning method for search result diversification in Twitter and demonstrated the effectiveness of the learning method by conducting an experiment. Zhu et al. (2017) presented a novel real-time personalized twitter search based on semantic expansion and quality model. More specifically, the authors developed a Boolean logic keyword filter to enhance the accuracy and built a tweet quality model trained by labeled data to improve ranking effectiveness. Their proposed method demonstrated superior performance against competitive baselines in a variety of metrics.

Bowler et al. (2015) investigated teens' perspectives on the quality and helpfulness of health information about eating disorders found on Yahoo! Answers, a Social Q&A site. The authors found that the teen participants made a distinction between health information in Social Q&A that is credible versus that which is helpful. In addition, they value health information that is not from a credible source if it addresses other needs. Furthermore, when making judgments about health information on the Web, the teen participants apply an array of heuristics related to information quality, opinion, communication style, emotional support and encouragement, guidance, personal experience, and professional expertise.

Bagdouri and Oard (2017) studied the possibility of answering the questions asked on Twitter using Yahoo! Answers, finding that two thirds of the answerable questions do have an excellent answer. The authors also found that searching in the title field of the old questions yields a significant improvement over a search in the concatenation of all the fields of a community question answering thread.

Ranking results refers to the process of quantifying the similarities between documents and queries. Social results ranking can be divided into two categories, depending on how social information is used (Bouadjenek et al, 2016). The first category uses social information to add a social relevance to the ranking process, whereas the second category utilizes it to personalize the search results. Social relevance refers to socially created information that characterizes a document from a point of view of interest, i.e., its popularity. Such social relevance can be added to the document ranking process. For example, He et al. (2014) proposed a method to predict the popularity of Web 2.0 items based on users' comments, and to incorporate this popularity into a ranking function. Their experimental results on three real world datasets—crawled from YouTube,

Flickr and Last.fm—showed that their proposed method consistently outperforms competitive baselines in several evaluation tasks.

Laniado, Eynard, and Colombetti (2007b) combined folksonomies from delicious.com with WordNet to execute semantically substantiated query expansions. Kim and Kim (2010b) proposed a structured folksonomy system in which queries can be expanded through tag control; equivalent, synonym or related tags are bound together, in order to improve the retrieval efficiency (recall and precision) of image data. Then, the authors evaluated the proposed system by comparing it to a tag–based system without tag control. Furthermore, the authors investigated which query expansion method is the most efficient in terms of retrieval performance. The experimental results showed that the recall, precision, and user satisfaction rates of the proposed system are statistically higher than the rates of the tag–based system, respectively. On the other hand, there are significant differences among the precision rates of query expansion methods but there are no significant differences among their recall rates.

Zhou et al. (2017) proposed a novel model to construct enriched user profiles with the help of an external corpus for personalized query expansion. Based on user profiles, the authors built two novel query expansion techniques. The results of an in–depth experimental evaluation showed that their approach outperforms traditional techniques, including existing non–personalized and personalized query expansion methods.

8.1.3 Social Recommendation

Exponential growth of information generated by social networks demands effective and scalable recommender systems in order to assist users in finding useful results (Jiang et al., 2014). Social recommendation has attracted a lot of attention recently in the research communities of information retrieval, machine learning, and data mining (Sonawane & Rokade, 2016). There are two main recommendation systems (Tang, Hu, & Liu, 2013); content based recommender systems that recommend items similar to the ones that the user has preferred in the past, and collaborative filtering based recommender systems that recommend items to the user based on the ratings or the behavior of other users or target users in the system (Duhan, 2018).

Ma et al. (2008, 2011) proposed a framework of social recom– mender systems that made use of social relation data, from which friendship information is exploited to regularize the user latent space. Ma (2013) examined a research problem on how to improve recommender systems using implicit social information. The author mentioned that implicit user and item social information, including similar and dissimilar relationships, can be employed to improve traditional recommendation methods, and that when comparing implicit social information with explicit social information, the performance of using implicit information is slightly worse.

Rawashdeh et al. (2013) proposed a tripartite graph-based approach to identify a list of tags that are personally tailored to a user's interests for a given item. Their proposed approach estimates the proximity between users and tags, and between items and tags based on the Katz measure, a path-ensemble based proximity measure, and thus discovers new triangle graphs that are likely to appear within a given folksonomy. The authors conducted experiments with the CiteULike and Last.fm datasets, demonstrating that the proposed method improves the recommendation performance and is effectiveness for both active taggers and cold-start taggers, compared to existing algorithms.

Jiang et al. (2014) suggested a novel social recommendation model utilizing two social contextual factors, individual preference and interpersonal influence. The authors conducted extensive experiments on both Facebook style bidirectional and Twitter style unidirectional social network datasets, and demonstrated that their method significantly outperforms the existing approaches.

To address the classical problems of natural language ambiguity caused by most content-based recommender systems, de Gemmis et al. (2015) classified semantic techniques into top–down and bottom–up approaches. The top–down approaches rely on the integration of external knowledge sources, such as taxonomies, thesauri or ontologies, for annotating items and representing user profiles in order to capture the semantics of the target user information needs. Conversely, the bottom–up approaches rely on a lightweight semantic representation under the assumption that the meaning of words depends on their use in large corpora of textual documents. The authors demonstrated how to make recommender systems aware of semantics to realize a new generation of content–based recommenders.

Godoy–Lorite et al. (2016) developed a rigorous probabilistic model that outperforms the leading approaches for recommendation and whose parameters can be fitted efficiently with an algorithm whose running time scales linearly with the size of the dataset. Zhao et al. (2016) presented a framework of online social recommendation from the viewpoint of a regularized online graph of user preference learning, which incorporates collaborative user–item relationships and item content features into a unified preference learning process. In a recent study, Duhan (2018) proposed an improved recommendation technique in order to address issues including accuracy, sparsity, and cold start, caused by collaborative filtering algorithms.

8.2 Case Studies: Social Indexing and Search

8.2.1 Overview

Taxonomies and folksonomies are used in descriptive metadata fields to support indexing and retrieval of digital content. Taxonomy can be defined as a controlled vocabulary (a set of controlled codes or terms usually designed by experts) that establishes hierarchical or associative relationships between terms (Smith, 2008a).

Although taxonomy have been used as information retrieval tools for a long time, taxonomies and controlled vocabularies suffer from various problems. First, there is a timeliness problem inherent in taxonomies and controlled vocabularies: taxonomies need to be agreed upon and codified into a classification (e.g., Dewey Decimal Classification) prior to their use by indexers (Peters, 2009). The timeliness problem can lead to a situation where classification structures are not compatible with current knowledge.

Second, indexers need to assign proper controlled terms to any subject materials. However, it is unreasonable to expect indexers to be experts in every field of knowledge (Steele, 2009). Such a problem would be even worse if controlled vocabularies were applied to digital libraries and the Web because of the large amounts of data collected therein.

Many users have started to index documents with their own keywords (tags) to make them retrievable. The indexing process is called social tagging, and the collection of tags used within one platform is called a folksonomy (Weller, 2007). Figure 8-3 shows the Pyramid image obtained from the Digital Collections at the UWM (University of Wisconsin–Milwaukee) libraries. The image is described by taxonomies (Thesaurus for Graphic Materials and Library of Congress Subject Headings) and folksonomies.

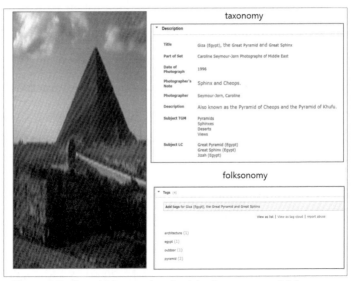

Figure 8-3. Pyramid image described by taxonomy and folksonomy

Folksonomies offer advantages related to cost–efficiency and feasibility for large data collections. Furthermore, folksonomies can be incredibly valuable for videos, images, and other media that do not have native text metadata (Smith, 2008a). Many studies suggest that folksonomies can be used as an efficient method for indexing and retrieving content as well as an alternative approach to solving the aforementioned problems associated with taxonomies and controlled vocabularies (Heymann, Koutrika, & Garcia−Molina, 2008; Yi, 2009; Geisler & Burns, 2007; Sharma & Elidrisi, 2008).

Heymann et al. (2008) examined the feasibility of using social tags as a source to augment Web search queries. They found that social tags were a promising resource for Web searching. Yi (2009) attempted to assess the indexing value of social tags in the context of an information−retrieval model using the Latent Semantic Indexing method. His study result showed the potential of using social tags as indexing terms for the DDC−based classification of tagged resources. Furthermore, Sharma and Elidrisi (2008) used video tags to classify YouTube videos into YouTube's predefined categories, and their results showed approximately 65% accuracy. Their study suggested that video tags could be used as metadata for videos.

However, folksonomies have their own problems, although many studies suggest that tags can be used efficiently as index terms. For example, the folksonomy−based approach lists tags without indicating relationships in flat name spaces, unlike the taxonomy−based approach, which displays words indicating relationships between them

(Matusiak, 2006). Thus, folksonomies do not include any vocabulary control; synonyms are not bound together and homonyms are not distinguished, which leads to a decrease in their retrieval effectiveness. The excitement over folksonomies being a better method of knowledge presentation and retrieval has given way to the realization that, without structure, they are less powerful than previously assumed (Peters, 2009). For example, the tags in some tagging systems have become non−navigable and are not even searchable due to the sheer volume or low quality of the tags.

In order to address these problems, we can combine the popularity and flexibility of folksonomies with the semantics and high−quality structures of taxonomies without sacrificing the features of folksonomies. To do so, we can build an ontology in which tags are related on the basis of paradigmatic (hierarchy, equivalence, or association) or syntagmatic (tag co−occurrences) relationships and then use it for query expansion or for tag recommendation when users assign tags.

Tag gardening, a concept introduced in a blog post by Governor (2006), is defined as any activity related to structuring and semantically disambiguating folksonomies in order to make them more productive and effective. The two main gardening activities are tag weeding and structuring: tag weeding is the process of removing bad tags (e.g., spam tags); tag structuring involves clustering tags based on paradigmatic or syntagmatic relationships or distinguishing homonymous tags (Peters & Weller, 2008). Smith (2008b) presented four aspects that play a role in tag gardening: 1) structuring folksonomies using semantic information; 2) the combination of automatic and manual tag editing; 3) leveraging user communities that contribute to structuring folksonomies; and 4) user-generated innovation such as developing new ways to apply tags.

8.2.2 Social Indexing of Videos

We analyzed YouTube tags to determine whether tags are useful for indexing and browsing videos (Kim, 2011). As sample data, we selected 300 videos with three or more tags from YouTube. Three thousand three hundred and twenty-five tags were extracted from the 300 sample videos. After removing 302 tags for reasons such as spam or spelling errors, 3,023 tags were finally selected as sample tags. For the tag analysis, we used a tag category framework. We detail the tag category framework, how to classify video tags using the framework, and the results of the analysis of video tag distribution in the following subsections.

1) Tag Category Framework
We created a tag category framework that has 5 categories and 17

subcategories (Table 8–1). We used the following studies to develop the category framework for the video tags. According to the previous studies (Panofsky, 1955; Eakins & Graham, 1999; Yoon, 2008), we assumed that people might interact with frames (images) of videos on three levels. On the first level, they observe the primitive features of a frame, such as color and shape (description). On the second level, they consider derived attributes like the presence of specific objects (analysis). Lastly, on the third level, they consider the semantic abstract attributes of the frame, such as the symbolic value (interpretation).

Thus, we included these three levels (categories) in our category framework. Furthermore, we added two categories, metadata (is–ness) and personal information (self–reference), to our tag category framework because after pre–analyzing sample video tags, we found that many tags were related to metadata (e.g., ownership/ source, creator names or IDs) or personal information (e.g., mystuff or mine) (Smith, 2008a). Next, after analyzing the Art and Architecture Thesaurus and the image categories suggested by Kim and Kim (2010a), Laine–Hermandez and Westman (2008), and Yoon (2008), we included 17 subcategories, such as color, objects, people, and metadata elements, to the category framework; each category has one or more subcategories

2) How to Classify Tags

According to the category framework, 3,023 tags (compound tags: 289 (9.6%), single-term tags: 2,734 (90.4%)) obtained from the 300 sample videos were analyzed. The following is an example of how to classify tags using the category framework. Figure 8-4 shows a picture of the video titled "How Benjamin Button got his face." The video deals with the special effects for the Academy Award-winning movie, The Curious Case of Benjamin Button. In this case, all of the sixteen tags assigned by users were single-word forms.

After checking the video content, we classified the 16 tags into 4 sub-categories: "TEDTalks," "TED," and "talks" are classified in the metadata elements subcategory of the metadata category because each of them is regarded as a part of the creator name (TEDTalks Director); seven tags including winner, Benjamin, and David are classified in the people subcategory of the analysis category; graphics is classified in the subject/theme subcategory of the interpretation category; and academy, award, Oscar, special, and effects are classified in the activities/events subcategory of the analysis category because they are regarded as parts of compound terms such as "special effects," "Academy Award," and "Oscar Award."

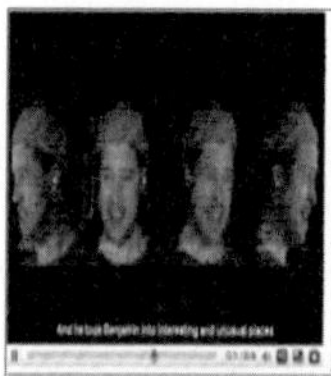

Tags: Ed Ulbrich TEDTalks TED talks academy award oscar winner special effects graphics Benjamin Button David Fincher	
creator	TEDtalksDirector
title	Ed Ulbrich: How Benjamin Button got his face

tag	subcategory	tag	subcategory
TEDTalks TED talks	metadata elements	graphics	subject/theme
Ed Ulbrich winner Benjamin Button	people	academy award oscar special effects	activities/ events

Figure 8-4. An example for classifying video tags (Kim, 2011)

3) The Analysis Result of Video Tag Distribution

According to the tag category framework and guidelines, 3,023 tags were analyzed; the most frequently used categories were analysis (51.74%), interpretation (29.94%), and metadata (16.44%), while the least frequently used categories were personal information (0.33%) and description (1.54%). On the other hand, the most frequently used subcategories were subject/theme (21.63%), objects (17.30%), metadata elements (16.44%), people (15.38%), and location (9.23%).

In the study by Kim and Kim (2009), which used the above-mentioned tag categorization system for classifying image tags, the most frequently used categories were analysis (43.24%), metadata (e.g., photography, camera, and group names; 30.43%), and interpretation (20.60%), while the least frequently used categories were personal information (0.44%) and description (5.30%). The most frequently assigned subcategories were metadata elements (30.40%), location (12.93%), objects (12.45%), and subject/theme (11.03%).

Thus, we can conclude that video tags seem to be closer to indexer–assigned terms than image tags, because YouTube videos have proportionally more tags describing video content (e.g., subject/theme–or object–related tags) than Flickr images have, so they are valuable as index terms. However, due to the lack of hierarchy or synonym control in folksonomies, a search for a specific term will only yield results for that term and not provide the full body of related terms that might be relevant to a user's information needs and goals (Gordon–Murnane, 2006). Therefore, folksonomies need tag clustering in which similar or related tags are summarized in order to be used efficiently in information retrieval.

Table 8-1. Tag distribution (videos) (Kim, 2011)

Description Category (1)			Analysis Category (2)		
Sub-categories	Sample Tags	Frequency (%)	Sub-categories	Sample Tags	Frequency (%)
color	green, color	11 (0.36)	people (person(s), social status, groups, etc.)	Johnson, teacher	465 (15.38)
texture/material	metal, mud	14 (0.46)	activities/events	interview, award	191 (6.32)
shape/ composition	round, plane	7 (0.23)	objects	book, car	523 (17.30)
number	2000, 2009	8 (0.26)	location	Africa, Paris	279 (9.23)
text (e.g., overlay text, letters on objects, etc.)	visions, voice	7 (0.23)	time	9/11, winter	51 (1.69)
			scenery	landscape, cityscape	55 (1.82)
Sub-total: 47 (1.54)			Sub-total: 1564 (51.74)		
Interpretation Category (3)			Metadata Category (4)		
Sub-categories	Sample Tags	Frequency (%)	Sub-categories	Sample Tags	Frequency (%)
abstract	old, love	226 (7.48)	metadata elements	genesmith (author), TEDTalks	497 (16.44)
atmosphere/ emotion	wow, great	6 (0.20)	Sub-total: 497(16.44)		
subject/theme	education, music	654 (21.63)	Personal Information Category (5)		
function	portrait, vacation	19 (0.63)	Sub-categories	Sample Tags	Frequency (%)
			personal information data	me, mine	10 (0.33)
Sub-total: 905 (29.94)			Sub-total: 10 (0.33)		
Grand Total: 3023 (100)					

8.2.3 Social Search of Videos

We investigated how effective is tag control through query expansion (tag gardening) in searching videos (Kim & Kim, 2010c; Kim, 2011). To do so, we designed a structured folksonomy-based system in which queries can be expanded through tag control; equivalent, synonymous, or related tags are bound together, in order to improve the retrieval effectiveness of videos. Then, we evaluate the proposed system by comparing it to a tag-based system without tag control.

1) System Design

We designed a structured folksonomy-based system. As sample data, we selected 300 videos with three or more tags from the YouTube site, because tags might be very useful for improving access to videos with limited textual metadata. The system enables users to expand their queries with synonymous or related tags. For this system, we created three word files (word-form, synonym, and related word files) for query expansion (Figure 8-5). The word-form file was created to summarize different forms of compounding, singular and plural forms, abbreviations and full-names, and multiple languages based on the use of Wikipedia or dictionaries. Then, the synonymous tag file was constructed to link tags via synonymous relationships based on the use of WordNet. Lastly, the related tag file was formed to link tags via syntagmatic relations based on the use of Flickr's related tags, which are generated based on co-occurrence analysis.

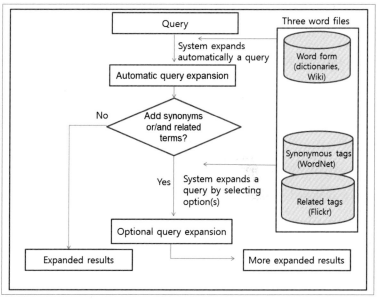

Figure 8-5. The procedure for expanding a query

We designed a system interface for the two systems. In the system, a user can submit a query to the system; the user can directly input a query or click the "alphabetical tag file" or "category tag file" and select tags from the list. Then, the query is automatically expanded using the word-form file. Next, the user can choose what type of relations (synonymous or/and related tags) should be used for expanding the query (see Figure 8–6).

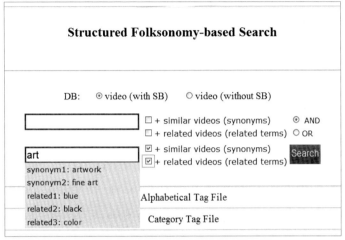

Figure 8-6. Structured folksonomy-based system interface (Kim and Kim, 2010c)

Then, we evaluated the proposed system by comparing it to a tag–based system without tag control, in terms of recall and precision rates. We used a t–test to compare the retrieval performance of the structured folksonomy–based system to that of tag–based system. As a result, the recall mean (0.63) of the folksonomy–based system was statistically higher that that (0.53) of the tag–based system. However, the precision mean (0.75) of the folksonomy–based system was not statistically higher that that (0.74) of the tag–based system.

References

Bagdouri, M., & Oard, D. W. (2017). Building bridges across social platforms: Answering Twitter questions with Yahoo! Answers. SIGIR. Tokyo, Japan.

Barzilay, R., & Elhadad, M. (1999). Using lexical chains for text summarization. Advances in automatic text summarization, 111~121.

Bouadjenek, M. R., Hacid, H., & Bouzeghoub, M. (2016). Social networks and information retrieval, how are they converging? A survey, a taxonomy and an analysis of social information retrieval approaches and platforms. Information Systems, 56, 1~18.

Bowler, L., Monahan, J., Jeng, W., Oh, J. S., & He, D. (2015). The quality and helpfulness of answers to eating disorder questions in Yahoo! Answers: Teens speak out. Proceedings of the Association for Information Science and Technology, 52(1), 1~10.

Cattuto, C., Benz, D., Hotho, A., & Stumme, G. (2008). Semantic analysis of tag similarity measures in collaborative tagging systems. arXiv preprint arXiv:0805.2045.

Choi, Y. (2014). Social indexing: A solution to the challenges of current information organization. In New directions in information organization (pp. 107~135). Emerald Group Publishing Limited.

Cui, C., Shen, J., Ma, J., & Lian, T. (2015, October). Social tag relevance estimation via ranking-oriented neighbour voting. In Proceedings of the 23rd ACM International Conference on Multimedia (pp. 895~898). ACM.

de Gemmis, M., Lops, P., Musto, C., Narducci, F., & Semeraro, G. (2015). Semantics−aware content−based recommender systems. In Recom− mender Systems Handbook (pp. 119~159). Springer US.

Dong, A., Zhang, R., Kolari, P., Bai, J., Diaz, F., Chang, Y., ... & Zha, H. (2010, April). Time is of the essence: Improving recency ranking using twitter data. In Proceedings of the 19th International Conference on World Wide Web (pp. 331~340). ACM.

Duhan, N. (2018). Collaborative Filtering−Based Recommender System. In ICT Based Innovations (pp. 195~202). Springer, Singapore.

Eakins, J. P., & Graham, M. E. (1999). Content−based image retrieval (JISC Technology Applications Programme Report 39).

Filho, F., Olson, G. M., & de Geus, P. L. (2010). Kolline: A task−oriented system for collaborative information seeking. In Proceedings of the 28th ACM International Conference on Design of Communication (pp. 89~94). ACM.

Furnas, G. W., Landauer, T. K., Gomez, L. M., & Dumais, S. T. (1987). The vocabulary problem in human−system communication. Comm− unications of the ACM, 30(11), 964~971.

Golub, K., Lykke, M., & Tudhope, D. (2014). Enhancing social tagging with automated keywords from the Dewey Decimal Classification. Journal of Documentation, 70(5), 801~828.

Governor, J. (2006). On the emergence of professional tag gardeners. Blog Post, 10.01.2006.

Geisler, G., & Burns, S. (2007). Tagging video: Conventions and strategies of the YouTube community [poster]. In E. M. Rasmussen, R. R. Larson,

E. G. Toms, & S. Sugimoto (Eds.), Proceedings of the Joint Conference on Digital Libraries (JCDL 2007) (p. 480). New York: ACM Press.

Godoy-Lorite, A., Guimerà, R., Moore, C., & Sales-Pardo, M. (2016). Accurate and scalable social recommendation using mixed-membership stochastic block models. Proceedings of the National Academy of Sciences, 113(50), 14207~14212.

Gordon-Murnane, L. (2006). Social bookmarking, folksonomies, and Web 2.0 Tools. Searcher, 14(6), 26~38.

Greenberg, J. (2001). Automatic query expansion via lexical-semantic relationships. Journal of the American Society for Information Science and Technology, 52(5), 402~415.

Greenaway, S., Thelwall, M., & Ding, Y. (2009, July). Tagging youtube-a classification of tagging practice on youtube. In 12th International Conference on Scientometrics and Informetrics, Rio De Janiro, Brazil.

Gupta, M., Li, R., Yin, Z., & Han, J. (2010). Survey on social tagging techniques. ACM Sigkdd Explorations Newsletter, 12(1), 58~72.

Hayman, S. (2007). Folksonomies and tagging: New developments in social bookmarking. Proceedings of the Ark Group Conference: Developing and Improving Classification Schemes. Sydney, Rydges World Square, 27~29 June.

He, X., Gao, M., Kan, M. Y., Liu, Y., & Sugiyama, K. (2014). Predicting the popularity of web 2.0 items based on user comments. In Proceedings of the 37th International ACM SIGIR Conference on Research & Development in Information Retrieval (pp. 233~242).

ACM.

Heckner, M., Neubauer, T., & Wolff, C. (2008).Tree, funny, to_read, Google: What are tags supposed to achieve? Proceedings of the 2008 ACM Workshop on Search in Social Media, Napa Valley, California, USA.

Heymann, P., Koutrika, G., & Garcia-Molina, H. (2008). Can social bookmarking improve web search? In M. Najork, A. Z. Broder, & S. Chakrabarti (Eds.), Proceedings of the International Conference on Web Search and Web Data Mining (pp. 195~206). New York: ACM Press.

Ingwersen, P. (2002, July). Cognitive perspectives of document representation. In Emerging frameworks and methods: CoLIS4: Proceedings of the Fourth International Conference on Conceptions of Library and Information Science, Seattle, WA, USA (pp. 285~300).

Jiang, M., Cui, P., Wang, F., Zhu, W., & Yang, S. (2014). Scalable recommendation with social contextual information. IEEE Transactions on Knowledge and Data Engineering, 26(11), 2789~2802.

Kennedy, L., Naaman, M., Ahern, S., Nair, R., & Rattenbury, T. (2007). How Flickr helps us make sense of the world: Context and content in community-contributed media collections. Proceedings of ACM Multimedia, 2007. Augsburg, Germany.

Kim, H. (2007). An experimental study on the effectiveness of storyboard surrogates in the meanings extraction of digital videos. Journal of Korean Society for Information Management, 24(4), 53~72.

Kim, H. (2011). Toward video semantic search based on a structured folksonomy. Journal of the American Society for Information Science, 62(3), 478~492.

Kim, H., & Kim, M. (2009). Investigating the end–user tagging behavior in Flickr. Journal of Korean Information Management, 40(2), 22~30.

Kim, H., & Kim, Y. (2010a). Toward a conceptual framework of key–frame extraction and storyboard display for video summarization. Journal of the American Society for Information Science and Technology, 61(5), 927~939.

Kim, H. & Kim, Y. (2010b). An experimental study on semantic searches for image data using structured social metadata. Journal of Korean Society for Library and Information Science, 44(1), 117~135.

Kim, H., & Kim, Y. (2010c). Semantic video search using tagsonomies. Proceedings of the American Society for Information Science and Technology, 47(1), 1~2.

Kim, H., & Kim, Y. (2016). Generic speech summarization of transcribed lecture videos: Using tags and their semantic relations. Journal of the Association for Information Science and Technology, 67(2), 366~379.

Knautz, K., Siebenlist, T., & Stock, W. G. (2010, July). Memose: Search engine for emotions in multimedia documents. In Proceedings of the 33rd International ACM SIGIR Conference on Research and Development in Information Retrieval (pp. 791~792). ACM.

Ko, Y., Kim, K., & Seo, J. (2003). Topic keyword identification for text summarization using lexical clustering. IEICE Transactions on Information and Systems, 86(9), 1695~1701.

Kolbitsch, J. (2007). WordFlickr: A solution to the vocabulary problem in

social tagging systems. Proceedings of I−MEDIA'07 and I−SEMAN
TICS'07, 77~84.

Kome, S. (2005). Hierarchical subject relationships in folksonomies.
University of North Carolina at Chapel Hill School of Information
and Library Science (MS Thesis, University of North Carolina at
Chapel Hill, 2005).

Laine−Hermandez, M., & Westman, S. (2008). Multifaceted image similarity
criteria as revealed by sorting tasks. Proceedings of the ASIST Annual
Meeting, 45, 1~14. Medford, NJ: Information Today.

Laniado, D., Eynard, D., & Colombetti, M. (2007a). A semantic tool to
support navigation in a folksonomy. Proceedings of the Eighteenth
Conference on Hypertext and Hypermedia (pp. 153~154). Manches−
ter, UK.

Laniado, D., Eynard, D., & Colombetti, M. (2007b). Using WordNet to
turn a folksonomy into a hierarchy of concepts. Proceedings of the
Italian Semantic Web Workshop−Semantic Web Application and
Perspectives (SWAP 2007) (pp. 192~201). Bari, Italy.

Lee, D. H., & Schleyer, T. (2012). Social tagging is no substitute for
controlled indexing: A comparison of Medical Subject Headings and
CiteULike tags assigned to 231,388 papers. Journal of the Association
for Information Science and Technology, 63(9), 1747~1757.

Lee, B., Fan, W., Squicciarini, A. C., Ge, S., & Huang, Y. (2014). The
relativity of privacy preservation based on social tagging. Information
Sciences, 288, 87~107.

Liu, F., & Lee, H. J. (2010). Use of social network information to enhance collaborative filtering performance. Expert Systems with Applications, 37(7), 4772~4778.

Ma, H. (2013). An experimental study on implicit social recommendation. In Proceedings of the 36th international ACM SIGIR Conference on Research and Development in Information Retrieval (pp. 73~82). ACM.

Ma, H., Yang, H., Lyu, M. R., & King, I. (2008, October). Sorec: Social recommendation using probabilistic matrix factorization. In Proceedings of the 17th ACM Conference on Information and Knowledge Management (pp. 931~940). ACM.

Ma, H., Zhou, D., Liu, C., Lyu, M. R., & King, I. (2011, February). Recommender systems with social regularization. In Proceedings of the Fourth ACM International Conference on Web Search and Data Mining (pp. 287~296). ACM.

Mandala, R., Tokunaga, T., & Tanoka, H. (1999). Combining multiple evidence from different types of thesaurus for query expansion. Proceedings of the 22nd Annual International ACMISIGIR Conference on Research and Development in Information Retrieval (pp. 191~197). Berkeley, CA: ACM.

Manning, C. D., Raghavan, P. & Schutze, H. (2008). Introduction to information retrieval. Cambridge University Press, Cambridge, England.

Mathes, A. (2004). Folksonomies—cooperative classification and communi-cation through shared metadata.

Matusiak, K. (2006). Towards user—centered indexing in digital image collections. OCLC Systems & Services: International digital library, 22(4), 283~298.

Melenhorst, M., Grootveld, M., van Setten, M., & Veenstra, M. (2008). Tag–based information retrieval of video content. In Proceedings of the 1st International Conference on Designing Interactive User Experiences for TV and Video (pp. 31~40). ACM.

Morris, M. R., Teevan, J., & Panovich, K. (2010, April). What do people ask their social networks, and why?: A survey study of status message q&a behavior. In Proceedings of the SIGCHI Conference on Human Factors in Computing Systems (pp. 1739~1748). ACM.

Morrison, J. (2008). Tagging and searching: Search retrieval effectiveness of folksonomies on the World Wide Web. Information Processing and Management, 44, 1562~1579.

Noruzi, A. (2007). Folksonomies: Why do we need controlled vocabulary?. Webology, 4(2).

Pan, X., He, S., Zhu, X., & Fu, Q. (2016). How users employ various popular tags to annotate resources in social tagging: An empirical study. Journal of the Association for Information Science and Technology, 67(5), 1121~1137.

Panofsky, E. (1955). Meaning in the visual arts: Meaning in and on art history. Garden City, NY: Doubleday.

Peters, I. (2009). Folksonomies: Indexing and retrieval in Web 2.0. Berlin: De Gruyter, Saur.

Peters, I., & Stock, W. G. (2007). Folksonomy and information retrieval. Proceedings of the ASIST Annual Meeting, 44 (pp. 1~18). Medford,

NJ: Information Today.

Peters, I. & Weller, K. (2008). Tag gardening for folksonomy enrichment and maintenance. Webology, 5(3).

Quintarelli, E. (2005). Folksonomies: Power to the people. Paper presented at the ISKO Italy UniMIB meeting, Milan, Italy.

Rawashdeh, M., Kim, H. N., Alja'am, J. M., & El Saddik, A. (2013). Folksonomy link prediction based on a tripartite graph for tag recommendation. Journal of Intelligent Information Systems, 40(2), 307~325.

Scale, M. S. (2008). Facebook as a social search engine and the implications for libraries in the twenty-first century. Library Hi Tech, 26(4), 540~556.

Sharma, A. S., & Elidrisi, M. (2008). Classification of multi-media content (videos on YouTube) using tags and focal points. Unpublished manuscript.

Shiri, A. (2009). An examination of social tagging interface features and functionalities: An analytical comparison. Online Information Review, 33(5), 901~919.

Siersdorfer, S., Pedro, J. S. & Sanderson, M. (2009). Automatic video tagging using content redundancy. 32nd ACM SIGIR Conference, Boston, MA: ACM.

Smith, G. (2008a). Tagging: People-powered metadata for the social Web. Berkeley: New Riders.

Smith, G. (2008b). Tagging: Emerging trends. Bulletin of the ASIST, 34(6), 14~17.

Smith-Yoshimura, K. (2011). Social metadata for libraries, archives and museums Part 1: Site Reviews.

Sonawane, P. M., & Rokade, S. M. (2016). Social recommendation system for real world online applications. International Journal, 1(8).

Song, Y., & Marchionini, G. (2007, April). Effects of audio and visual surrogates for making sense of digital video. In Proceedings of the SIGCHI Conference on Human Factors in Computing Systems (pp. 867~876). ACM.

Specia, L., & Motta, E. (2007). Integrating folksonomies with the semantic web. The semantic web: Research and applications, 624~639.

Steele, T. (2009). The new cooperative cataloging. Library Hi Tech, 27(1), 68~77.

Syn, S. Y., & Spring, M. B. (2013). Finding subject terms for classificatory metadata from user-generated social tags. Journal of the Association for Information Science and Technology, 64(5), 964~980.

Tang, J., Hu, X., & Liu, H. (2013). Social recommendation: A review. Social Network Analysis and Mining, 3(4), 1113~1133.

Teevan, J., Ramage, D., & Morris, M. R. (2011, February). # TwitterSearch: A comparison of microblog search and web search. In Proceedings of the fourth ACM International Conference on Web Search and Data Mining (pp. 35~44). ACM.

Trant, J. (2006). Exploring the potential for social tagging and folksonomy in art museums: Proof of concept. New Review of Hypermedia and Multimedia, 12(1), 83~105.

Wang, Y., Luo, Z., & Yu, Y. (2016, June). Learning for search results diversification in Twitter. In International Conference on Web-Age Information Management (pp. 251~264). Springer, Cham.

Weller, K. (2007). Folksonomies and ontologies: Two new players in indexing and knowledge representation. In H. Jezzard (Ed.), Applying Web 2.0. Innovation, Impact and Implementation: Online Information 2007 Conference Proceedings (pp. 108~115). London: Incisive Media.

Yang, M., & Marchionini, G. (2004). Exploring users' video relevance criteria: A pilot study. Proceedings of the ASIST Annual Meeting, 41 (pp. 229~238). Medford, NJ: Information Today.

Yi, K., & Chan, L. M. (2009). Linking folksonomy to Library of Congress subject headings: An exploratory study. Journal of Documentation, 65(6), 872~900.

Yi, K. (2009). A study of evaluating the value of social tags as indexing terms. Proceedings of the Sixth International Conference on Knowledge Management [CD-ROM]. Singapore: World Scientific Publishing. Hong Kong.

Yoo, D., & Suh, Y. (2010). User-categorized tags to build a structured folksonomy. Proceedings of the 2nd International Conference on Communication Software and Networks (ICCSN 2010) (pp. 160~164). Singapore, Feb. 26~28, 2010.

Yoon, J. (2008). Searching for an image conveying connotative meanings: An exploratory cross-cultural study. Library & Information Science Research, 30(4), 312~318.

Zhao, Z., Lu, H., Cai, D., He, X., & Zhuang, Y. (2016). User preference learning for online social recommendation. IEEE Transactions on Knowledge and Data Engineering, 28(9), 2522~2534.

Zhou, D., Wu, X., Zhao, W., Lawless, S., & Liu, J. (2017). Query expansion with enriched user profiles for personalized search utilizing folksonomy data. IEEE Transactions on Knowledge and Data Engineering.

Zhu, X., Huang, J., Zhou, B., Li, A., & Jia, Y. (2017). Real-time personalized twitter search based on semantic expansion and quality model. Neurocomputing, 254, 13~21.

Chapter 9

Evaluation

9. Evaluation

In this chapter, we describe multimedia retrieval focusing on the approach and performance evaluation. Then, in order to verify the feasibility of our research results, we evaluate the effectiveness of our proposed social summarization method and video summarization using EEG/ERP Techniques, which were introduced in Chapters 6 and 7, respectively.

9.1 Multimedia Retrieval: Approach and Performance Evaluation

In order to search relevant multimedia items, we can leverage two different methods, such as metadata−based multimedia retrieval (e.g., using traditional metadata elements, titles and keywords) and content−based multimedia retrieval (CBMR) (e.g., using automatic feature and object recognition) (Amato et al., 2015; Colace et al., 2015). For a metadata−based image search, an image is manually annotated by metadata elements. This approach has two main disadvantages; a considerable level of human effort is required for manual annotation and the annotation inaccuracy due to the subjectivity of human perception.

To overcome the disadvantages in metadata–based retrieval systems, the content–based image retrieval (CBIR) using color, texture, and shapes was introduced. The most significant problem in CBIR appears to be the semantic gap problem (Jain & Sinha, 2010). The semantic gap comes from the lack of coincidence between the information extracted from the visual data and the user's interpretation for the same data (Smeulders et al., 2000).

Cox et al. (2000) demonstrated that CBIR can be improved through relevance feedback by involving the user in the search loop. However, relevance feedback can lead to a context trap, where a user specifies the context so strictly that he or she can only exploit a limited area of information space (Glowacka & Hore, 2014). In order to avoid the context trap, Suditu and Fleuret (2012) tried to combine exploration and exploitation strategies with relevance feedback. There are three types of image search (Hore et al., 2015): 1) a targeted search, looking for a specific image; 2) a category search, looking for a category of an image; and 3) an open–ended search, finding a story, i.e., a set of images that provide a theme for a document.

Several CBIR search engines have been developed, such as QBIC

(Query By Image Content), Pixolution (http://pixolution.org/), Google Image Search, and TinEye. IBM developed the QBIC system in the early 1990s, and the system makes use of all three primary visual features such as color, shape, and texture. The QBIC system incorporates many query tools such as color histogram selection, query by image example, and sketch tools (Flickner et al., 1997; Tseng, 2012). The QBIC system also allows for the textual representations of images, which can then be used for querying as well. As can be observed in Figure 9−1, the Pixolution Web interface combines key−based and content−based queries to adopt users' notions of interestingness. Figure 9−1 indicates the screenshot shown when selecting the most relevant image from the images retrieved by the keyword query "banana."

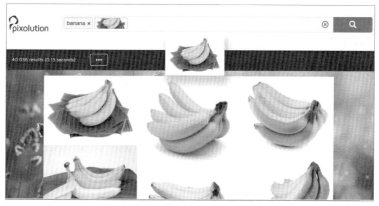

Figure 9-1. Pixolution search result

Recently, Al-Quraan, Nusir, and Abuata (2017) compared the performance of three image search engines (Yahoo, Ask, and Google) to answer Arabic text queries, and found that Google had the best retrieval effectiveness. Measures of image retrieval can be defined in terms of precision and recall. However, there are also rank-based measures including 1) the rank of the first relevant image (Rank1), as indicated by its name, the value of Rank1 expresses the position of the first relevant image in an ordered result list; and 2) the average rank of relevant images (Grubinger, 2007; Gargi & Kasturi, 1999).

Cheng and Shen (2016) introduced the construction of a large-scale landmark image dataset, which contained various kinds of textual features (e.g., tags) and six types of visual features (e.g., color histogram). Based on the dataset, they conducted a set of experimental studies on landmark search using visual and textual features. The result disclosed the weakness of content-based landmark search; both search accuracy and efficiency need to be improved. It also demonstrated that text-based retrieval methods for landmark image search using social tags can achieve good search performance and the combination of content-based and text-based retrieval methods can improve search accuracy in general.

ImageCLEF, which began as a part of the CLEF (Cross Language Evaluation Forum) in 2003 (Peters et al., 2003; Grubinger, 2007), is known to be the most significant and influential evaluation event for image retrieval. The goals of ImageCLEF are to investigate the effectiveness of combining text and images for retrieval, to collect and provide resources for benchmarking image retrieval systems, and to evaluate the systems in a multilingual environment. To achieve these goals, during the first four years, ImageCLEF provided three tasks, such as ad—hoc retrieval, object recognition, and interactive evaluation. Here, the ad—hoc retrieval task is to find as many relevant images as possible from a given document collection when textual statements and/or sample images were given, whereas the interactive evaluation task is concerned with the study of cross—language image retrieval from a user—centered perspective. Recently, ImageCLEF (Villegas et al., 2015; Schaer et al., 2016) provided various tasks including (1) automatic concept annotation, localization and sentence description generation for general images; (2) identification, multi—label classification and separation of compound figures from biomedical literature; (3) automatic annotation of general web images; and (4) retrieval from collections of scanned handwritten documents.

9.2 The Evaluation of Social Summaries Created by the Social Summarization Method

We compared the effectiveness and uniqueness of the tag–based framework with those of the LSA method (Kim & Kim, 2016). To accomplish this, we first used intrinsic evaluation, where the quality of a generated summary is directly judged through its analysis. We then used extrinsic evaluation, where a summary is assessed in terms of the extent to which it provides accurate information about a speech video when a user determines its relevance to his or her information need. Through these evaluations, we intended to ascertain whether the results of the intrinsic measure correlated with those of the extrinsic measure.

9.2.1 Method

We explain the application of both the intrinsic and extrinsic evaluation methods to our study as follows:

1) Intrinsic evaluation

We compared the summaries generated by the tag–based and LSA methods with reference summaries generated by human coders. A reference summary for each speech was constructed by two coders—

the authors of this study. We chose key sentences that describe the topic of the speech, its purpose, argument/conclusion, background, and method/activity (Kamal & Rubin, 2010). We used the ROUGE-1 (unigram co-occurrence) measure and the f-measure for a comparison. The ROUGE-1 measure has been shown to be highly effective for evaluating very short summaries (Lin, 2004). It is calculated as the number of overlapping single terms between a computer-generated evaluation summary and a reference summary. However, in our study, it was measured as the similarity between them using term vectors and cosine similarity.

2) Extrinsic evaluation

For extrinsic evaluation, we used summaries in Korean and collected data from participants through questionnaires in Korean. We selected 40 undergraduate students majoring in the humanities and social science, and randomly divided them into two groups (groups I and II). Furthermore, we randomly selected 30 videos from a sample of 70 videos and used their summaries (30 tag-and 30 LSA-based summaries) as sample data. The 40 participants were generally not fluent in English and had not been trained for abstracting documents. We thus translated 30 tag-and 30 LSA-based summaries from English to Korean using transcripts in Korean obtained from the TED talk website (http://www.ted.com/Open TranslationProject).

We then compared the 30 tag–and the 30 LSA–based Korean summaries in terms of the degree to which participants understood the content of the lecture speeches when reading the corresponding summaries generated by our proposed method and those generated by the LSA method.

For this, we formulated two questionnaires (A and B). Questionnaire A includes questions asking each participant in group I to carry out an open–ended summarizing task describing the content of the speech (e.g., topic) in three Korean sentences after reading each of 30 Korean summaries extracted by the tag–based framework. Questionnaire B includes questions asking each participant in group II to conduct the same task after reading each of 30 Korean summaries extracted by the LSA method. We asked for a three–sentence description for each speech because a few participants tended to write extremely short one–sentence descriptions.

The accuracy of each description was evaluated on a 30–point scale by the two coders (the scale ranged from 0–30, where zero indicates that the description is completely wrong, and 30 represents a completely correct summary). The correlation between the scores provided by the two coders was highly correlated (Pearson $r = 0.64$, p [$= 0.00$] < 0.01). If the discrepancy was greater than or equal to

10, the two original coders and a new coder were asked to discuss the reasons for the discrepancy and reach a consensus on a final score. If the discrepancy between two assigned scores was less than 10, the average was used as the final score.

9.2.2 Results

We describe here the evaluation results obtained from the two methods.

1) Intrinsic Evaluation: We conducted a one-way analysis of variance (ANOVA) to determine whether the mean rates for the f-measure and the ROUGE-1 measure are statistically different among the four methods: the proposed tag-based method, the tag method, the LSA (avg. = 12.1) method, and the LSA (avg. = 81.4) method. For the tag method, we utilized only original tags and title words (excluding expanded tags and the semantic relations between tags), and then used cosine similarity to compute the similarity of a sentence and a tag set. The ANOVA test confirmed a statistically significant difference among the four methods in terms of f-measure rates $(F = 2.98, p\ [= 0.03] < 0.05)$ and ROUGE-1 measures $(F = 2.67, p\ [= 0.048] < 0.05)$.

Table 9-1. Descriptive information for f-measure and ROUGE-1 measure (Kim & Kim, 2016).

| Method / Measure | Mean and S.D. of f-measure and ROUGE-1 measure | | | |
	Tag-based method ($\alpha = 0.5$ and $\beta = 0.5$)	Tag method (using original tags and title words)	LSA (avg. = 81.4) method	LSA (avg. = 12.1) method
f−measure	0.30 (0.24)	0.28 (0.16)	0.21 (0.22)	0.23 (0.23)
ROUGE−1 measure	0.28 (0.13)	0.27 (0.11)	0.22 (0.15)	0.25 (0.15)

2) Extrinsic evaluation: We compared the accuracy scores of the summaries produced by the proposed method to those by the LSA (avg. = 81.4) method with a t−test, in order to verify whether the intrinsic test results, which showed that our proposed method is better than the LSA method, agree with the extrinsic evaluation. As expected, the grand mean accuracy score of the proposed method (19.33) was found to be higher than that of the LSA (avg. = 81.4) method (12.55) ($t = 8.10$, p [= 0.02] < 0.05) with a statistically significant difference.

Table 9-2. Example summaries

No.	Tag (weighted value)	LSA (weighted value)
V4	S3: Learning English With Mister Duncan/Lesson Forty–three. (0.27) S20: The word 'superstition' comes from the ancient language of Latin and literally means 'stand over.' (0.27) S65: In China, the number 4 is considered unlucky because in the Chinese language four has a similar sound to the Chinese word for death. (0.27)	S37: Good superstition/Touching something made of wood is supposed to bring you good luck in the future. (0.16) S18: Good Fortune or Good Luck. (0.12) S49: Keeping a rabbit foot in your pocket is supposed to bring good luck. (0.11)
V20	S72: Last May, China in the Sichuan province had a terrible earthquake, 79 magnitude, massive destruction in a wide area, as the Richter Scale has it. (0.37) S165: We saw some of the most imaginative use of social media during the Obama campaign. (0.36) S83: The BBC got their first wind of the Chinese earthquake from Twitter. (0.27) S85: The last time China had a quake of that magnitude it took them three months to admit that it had happened. (0.26) S101: People began to figure out, in the Sichuan Provence, that the reason so many school buildings had collapsed, because tragically the earthquake happened during a school day, the reason so many school buildings collapsed is	S194: Media, the media landscape that we knew, as familiar as it was, as easy conceptually as it was to deal with the idea that professionals broadcast messages to amateurs, is increasingly slipping away. (0.11) S38: The media that is good at creating conversations is no good at creating groups. (0.09) S118: And the Great Firewall of China is a set of observation points that assume that media is produced by professionals, it mostly comes in from the outside world, it comes in relatively sparse chunks, and it comes in relatively slowly. (0.08) S53: The second big change is that as all media gets digitized the Internet also becomes the mode of carriage for all other media. (0.07) S62: Every time a new consumer joins this

No.	Tag (weighted value)	LSA (weighted value)
	that corrupt officials had taken bribes to allow those building to be built to less than code. (0.23) S191: Nobody in the Obama campaign had ever tried to hide the group or make it harder to join, to deny its existence, to delete it, to take to off the site. (0.23)	media landscape a new producer joins as well. (0.07) S35: This is the media landscape as we knew it in the 20th century. (0.06)
V66	S13: And so throughout the whole book, Al Gore will walk you through and explain the photos. (0.19) S31: And so you can start reading on your iPad in your living room and then pick up where you left off on the iPhone. (0.19) S2: It's called "Our Choice" and the author is Al Gore. (0.16)	S1: So for the past year and a half, my team at Push Pop Press and Charlie Melcher and Melcher Media have been working on creating the first feature–length interactive book. (0.39) S13: And so throughout the whole book, Al Gore will walk you through and explain the photos. (0.18) S6: As the globe spins, we can see our location, and we can open the book and swipe through the chapters to browse the book. (0.17)

For example, the summary of V4 (its main theme was "English lesson about superstition by Mister Duncan") generated by the LSA method consisted of three sentences. These sentences were selected based on the phrase "good luck," which was used to describe superstition. In the video, superstition was employed as a subject for the English lesson. Therefore, the participants found it difficult to identify the content of V4 from the summary.

On the other hand, it was relatively easy for them to grasp the main topic from the summary generated by our proposed method. This is because this summary included a sentence (S3) describing its main theme as well as one (S20) defining the subject (superstition) used for the English lesson. As expected, the accuracy score of the summary generated by the proposed method (20.71) was higher than that of the summary generated by the LSA−based method (5.0) by a statistically significant difference.

Further, although the summary of V20 generated by the LSA method describes the main concept—"media landscape"—it was found to be similar to or worse in terms of its quality (11.82) in comparison with the summary generated by the proposed method (16.0), which explains how social media is used for election and news delivery.

The summary of V66 by the LSA method describes the main concept—"interactive book"—whereas that generated by the proposed method does not include any sentence related to this concept. As a result, unlike in previous cases, the summary generated using the LSA method (15.0) proved superior to that generated by our tag−based method (6.25), with a statistically significant difference.

9.2.3 Discussion

Based on our findings from the intrinsic and extrinsic evaluations, we conclude that the quality of the tag–based summaries is as good as or better than the LSA–based summaries for most speech videos. The test results have several theoretical and practical implications for speech video summarization, retrieval, and browsing.

A closer examination of the tagging process assumed by the proposed method gives us an opportunity to compare it with the summarization processes assumed by other methods. Tags in the proposed method are conceived of as a reflection of an author's or a viewer's conceptual model of information, and are used as authentic representations of his or her language (Quintarelli, 2005; Peters & Stock, 2007).

Other approaches use extracting algorithms for the summarizing process. For example, the LSA method uses the context of an input document to extract key concepts based on syntagmatic (word co–occurrences) relationships between words, but disregards multiple meanings of a word (polysemy), its syntactic relations, and information about word order (Jorge–Botana, León, Olmos, & Hassan–Montero, 2010; Ozsoy et al., 2011). Moreover, it reduces the size of an input matrix by identifying and selecting content words or phrases from a given transcript to improve its effectiveness.

Both methods are similar in that they tend to select sentences with semantically related key concepts. However, each has different procedures for obtaining key concepts. Our proposed method uses viewer–assigned tags directly as key concepts, whereas the LSA method carries out several tasks, such as choosing topic–related words or phrases from an input text and conducting complex computations, for this purpose. Further, we observe that the proposed method enables the generation of summaries based on several concepts. Thus, the summary may present a comprehensive picture of a speech. On the other hand, the LSA method tends to generate the summary by focusing on one or two concepts, enabling it to include detailed information about the concept(s).

9.3 The Evaluation of EEG-based Key Shot Extraction Algorithm

We proposed an EEG–based key shot extraction algorithm (refer Chapter 7.4 "Video Summarization Using EEG/ERP Techniques" for more details). In order to evaluate this algorithm, first, we conduct a discriminant analysis to determine which ERP components and electrodes are important in distinguishing between relevant and irrelevant shots, and then to evaluate our proposed method using the discriminant function from the discriminant analysis. We then evaluate our proposed method using the intrinsic evaluation method.

9.3.1 Discriminant Analysis: Variable Selection and Evaluation

We used linear discriminant analysis to classify video shots as relevant or irrelevant because discriminant analyses have been successfully used in EEG signal classification (Ince, Goksu, Tewfik, & Arica, 2009; Xu et al., 2014). A stepwise discriminant analysis was used to find which ERP components and electrodes are important for distinguishing relevant or irrelevant shots. When performing this type of analysis, one can split a dataset into a training set and a test set when there are a sufficiently large number of cases. The training set is used to build a discriminant function and the test set is used to estimate predictability. We only had 21 cases in this study. Therefore, a cross−validation method, which is used when no test set is available, was utilized wherein each case was classified by the discriminate function derived from all cases (20 cases) other than that case (itself).

In our example, the dependent variable was the category (1= irrelevant, 3=relevant), and the independent variables were the means of the minimum or maximum amplitudes of the 18 electrodes (8 electrodes for N400 and 10 electrodes for P600) with significant differences between the two conditions as assessed by the previous t−tests (refer Chapter 7.4.4 for more details).

An unstandardized stepwise discriminant analysis identified two electrodes (Cz and CPz) in the P600 time window as the independent variables permitting discrimination between relevant shots and irrelevant shots. These two discriminating variables were used in the following discriminant function:

$$D = 0.859 \cdot Cz\_P600 - 0.572 \cdot CPz\_P600 - 0.565$$

where Cz_P600 represents the Cz electrode in the P600 time window, and the coefficient, 0.859 before the variable represents its contribution proportion to the discriminating power of the function. Table 9–3 shows the results of the classification as a simple summary of numbers and percentages of cases classified correctly and incorrectly.

Table 9-3. Classification results

			Predicted group membership		Total
			Irrelevant	Relevant	
Original	Count	1	18	3	21
		3	1	20	21
	%	1	85.7	14.3	100.0
		3	4.8	95.2	100.0
Cross−validated[a]	Count	1	18	3	21
		3	2	19	21
	%	1	85.7	14.3	100.0
		3	9.5	90.5	100.0

[a] Cross validation is done only for those cases in the analysis.

The discriminant function was able to correctly classify 90.5% of the original grouped cases and 88.1% of the cross–validated grouped cases. We think that the classification success rates are so high because, as mentioned before, we excluded 27 unmatched epochs (cases) for this discrimination analysis.

9.3.2 Intrinsic Evaluation

We compared the video summaries generated by the proposed and mechanical methods with reference summaries (ground truth) generated by human coders.

1) Reference Summary Generation

Ground truth comparison–based evaluation metrics are widely used to evaluate the effectiveness of video summarization methods. In this study, before constructing a ground truth database, we collected materials related to the test videos including their video skims, which comprise key shots extracted from the test videos. Next, three multimedia experts manually generated ground truth summaries using the following guidelines (Over, Smeaton, & Awad, 2008; Kim & Kim, 2010):

i) The task of the ground truth generator is to select desired shots including shots that appear in the video skim if a test video has its video skim. They then uniquely identify each by noting an object (animate or inanimate) or event (i.e., one or more objects involved in some action) occurring in the shot. Each shot typically lasts 2–5 s. The number of shots will vary per video.

ii) From a set of candidate shots, first select shots that include frames having overlay text, schematics, or symbols (T–frames). If there are redundant shots that have T–frames, then select those shots that appear in the opening or ending sequences of a video (or a corresponding video skim).

iii) The object/event cue for each desired shot should be as simple as possible while still identifying the shot uniquely within the video. Hence, select shots that include object frames (O–frames), action/event frames (E–frames), and person frames (P–frames) that indicate the central theme of a video. If multiple redundant shots include P–frames, either select the shot that appears in a corresponding video skim or select the shot that includes the greatest number of key individuals. We do not specify an order for the selection of shots of each type, because it is dependent on video genre; some genres will require an emphasis on a certain person or event.

2) Generating ERP-based Video Skims

We describe an ERP model, and then explain the process of generating video skims using the ERP model.

(i) ERP Model

The ERP model is designed to generate summaries by choosing the most relevant shots. In the extractive summarization, the score of a shot S_i in the kth iteration of the ERP is calculated as follows:

$$ERP = Dvalue\ (S_i)$$

where Dvalue (S_i) is the average discriminant score for the topical relevance of a shot S_i. The average discriminant score is computed through the above−mentioned discriminant function $(D = 0.859 \cdot Cz\_P600 - 0.572 \cdot CPz\_P600 - 0.565)$. The shots with the highest ERP scores will be iteratively chosen in the summary until it reaches a predefined size.

(ii) Construction of ERP-based Video Skims

We constructed an ERP−based video skim (Figure 9−2) for Video 2 that consists of three shots. We used f−measure to evaluate our video summarization method. The f−measure is the average of the precision and recall; a high f−measure indicates that both the precision and recall have high values. We compared the f–measure scores of the video skims for Video 2 generated by our proposed ERP–based, and mechanical methods with ground truth skims (see Figure 9–2 and Table 9–4). The f–measure score (0.67) of its ERP–based summary is higher, compared with that (0.33) of its mechanical method–based summary, which was employed to extract every n^{th} shot mechanically, where n equals the total number of shots of a video divided by the number of key shots we need.

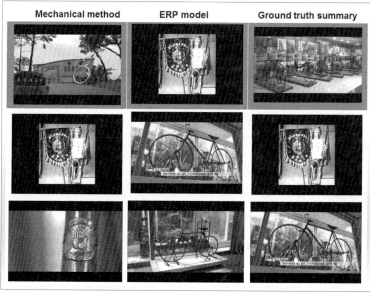

Figure 9-2. Video summaries for Video 2

3) Evaluation Based on f-measure, Redundant Shots, and Junk
   Shots

We compared the video skims generated by our proposed and mechanical methods with ground truth summaries. As mentioned above, as evaluation criteria, we used the f-measure, which is a widely employed metric for evaluating video summarization and video retrieval tasks (Dumont & Merialdo, 2010). We also used the amount of redundancy present, and the proportion of junk shots present in the summary. The scores for a lack of redundancy (RE) and lack of junk shots (JU) were normalized within a range of 1 to 5; a score close to 5 denotes that the video summary includes minimum redundancy and minimum junk shots, whereas a score close to 1 denotes that the video summary contains redundant and junk shots.

Table 9-4. The evaluation results of the ERP-based, and mechanical methods

no.	Title	ERP method			Mechanical method (SBD)		
		F	RE	JU	F	RE	JU
1	Discovering glasses	0.67	5	5	0.33	5	5
2	Uhm's bicycle	0.67	4	5	0.33	5	4
3	Two pocket watches	0.76	4	5	0.36	5	5
	Average	0.70	4.3	5	0.34	5	4.7

As shown in Table 9–4, the f–measure scores of the proposed ERP method are higher than those of the mechanical method, whereas the scores for a lack of junk shots of our proposed ERP method are also higher than those of the mechanical method. Conversely, the scores for a lack of redundant shots of our proposed ERP method are lower than those of the mechanical method. If we check the video skim for Video 2, which was constructed using the ERP model, we notice two visually similar shots (the second and third shots) (see Figure 9–2).

Therefore, in order to generate diverse summaries, our proposed ERP method can adapt the MMR method (see Chapter 6.4) that permits more diverse video summaries by choosing shots with high discriminant scores for topical relevance that are computed through the analysis of viewers' ERP data while ensuring that there is minimal duplication of shots already chosen for a video skim.

References

Al–Quraan, L., Nusir, S., & Abuata, B. (2017). Content based web image search engine evaluation using Arabic text queries. International Journal of Applied Research on Information Technology and Computing, 8(2), 125~140.

Amato, F., Greco, L., Persia, F., Poccia, S. R., & De Santo, A. (2015).

Content−based multimedia retrieval. In Data Management in Pervasive Systems (pp. 291~310). Springer International Publishing.

Cheng, Z., & Shen, J. (2016). On very large scale test collection for landmark image search benchmarking. Signal Processing, 124, 13~26.

Colace, F., De Santo, M., Moscato, V., Picariello, A., Schreiber, F. A., & Tanca, L. (2015). Data Management in Pervasive Systems. Springer.

Cox, I. J., Miller, M. L., Minka, T. P., Papathomas, T. V., & Yianilos, P. N. (2000). The Bayesian image retrieval system, PicHunter: Theory, implementation, and psychophysical experiments. IEEE transactions on image processing, 9(1), 20~37.

Flickner, M., Sawhney, H., Niblack, W., Ashley, J., Huang, Q., Dom, B., et al. (1997). Query by image and video content: The QBIC system. In M. T. Maybury (Ed.), Intelligent Multimedia Information Retrieval. California, USA: MIT Press.

Gargi, U., & Kasturi, R. (1999). Image database querying using a multi−scale localized color representation. In Content−based Access of Image and Video Libraries, 1999.(CBAIVL'99) Proceedings. IEEE Workshop on (pp. 28~32). IEEE.

Głowacka, D., & Hore, S. (2014). Balancing exploration−exploitation in image retrieval. Proceedings of the 22nd Conference on User Modeling, Adaptation, and Personalization. Aalborg, Denmark.

Grubinger, M. (2007). Analysis and evaluation of visual information systems performance. PhD thesis, Victoria University.

Hore, S., Glowacka, D., Kosunen, I., Athukorala, K., & Jacucci, G. (2015, August). FutureView: Enhancing exploratory image search. In IntRS@

RecSys (pp. 37~40).

Ince, N. F., Goksu, F., Tewfik, A. H., & Arica, S. (2009). Adapting subject specific motor imagery EEG patterns in space—time—frequency for a brain computer interface. Biomedical Signal Processing and Control, 4(3), 236~246.

Jain, R., & Sinha, P. (2010). Content without context is meaningless. Proceedings of the International Conference on Multimedia (MM '10) (pp. 1259~1268). NY, USA: ACM.

Jorge—Botana, G., León, J., Olmos, R., & Hassan—Montero, Y. (2010). Visualizing polysemy using LSA and the predication algorithm. Journal of the American Society for Information Science and Technology, 61(8), 1706~1724.

Kamal, A. M., & Rubin, V. L. (2010). Human abstracts, machine summaries, cyborg solutions?. Proceedings of the Association for Information Science and Technology, 47(1), 1~2.

Kim, H., & Kim, Y. (2010). Toward a conceptual framework of key-frame extraction and storyboard display for video summarization. Journal of the Association for Information Science and Technology, 61(5), 927~939.

Kim, H., & Kim, Y. (2016). Generic speech summarization of transcribed lecture videos: Using tags and their semantic relations. Journal of the Association for Information Science and Technology, 67(2), 366~379.

Lin, C. (2004). ROUGE: A package for automatic evaluation of summaries. In Proceedings of the Workshop on Text Summarization Branches Out (WAS 2004), Barcelona, Spain.

Nieuwenhuysen, P. (2015). Search by image through the Internet: An additional method to find information. In: International Conference on Libraries (ICOL) 2015, 25~26 August 2015, Vistana Hotel, Pulau Pinang.

Over, P., Smeaton, A. F., & Awad, G. (2008). The TRECVid 2008 BBC rushes summarization evaluation. In Proceedings of the 2nd ACM TRECVid Video Summarization Workshop (pp. 1~20). ACM.

Ozsoy, M., Alpaslan, F., & Cicekli, I. (2011). Text summarization using latent semantic analysis. Journal of Information Science, 37(4), 405~417.

Ozsoy, M., Cicekli, I., & Alpaslan, F. (2010). Text summarization of Turkish texts using latent semantic analysis. In Proceedings of the 23rd International Conference on Computational Linguistics (pp. 869~876).

Peters, C., Braschler, M., Gonzalo, J., & Kluck, M. (2003). Comparative evaluation of multilingual information access systems: Fourth Workshop of the Cross-Language Evaluation Forum (CLEF 2003), volume 3237 of Lecture Notes in Computer Science (LNCS). Springer.

Peters, I., & Stock, W.G. (2007). Folksonomy and information retrieval. Proceedings of American Society for Information Science and Technology Annual Meeting, 44(1), 1~18.

Quintarelli, E. (2005). Folksonomies: Power to the people. Paper presented at the ISKO Italy UniMIB meeting, Milan, Italy.

Schaer, R. et al. (2016, August). General overview of ImageCLEF at the

CLEF 2016 Labs. In Experimental IR Meets Multilinguality, Multimodality, and Interaction: 7th International Conference of the CLEF Association, CLEF 2016, Évora, Portugal, September 5~8, 2016, Proceedings (Vol. 9822, p. 267). Springer.

Smeulders, A., Worring, M., Santini, S., Gupta, A., & Jain, R. (2000). Content-based image retrieval at the end of the early years. IEEE Transactions on Pattern Analysis and Machine Intelligence, 22(12), 1349~1380.

Suditu, N., & Fleuret, F. (2012, October). Iterative relevance feedback with adaptive exploration/exploitation trade-off. Proceedings of the 21st ACM International Conference on Information and Knowledge Management (pp. 1323~1331). ACM.

Tseng, L. C. (2012). Modelling users' contextual querying behaviour for web image searching (Doctoral dissertation, Queensland University of Technology).

Villegas, M. et al. (2015). General overview of ImageCLEF at the CLEF 2015 labs. In: Working Notes of CLEF 2015. Lecture Notes in Computer Science. Springer International Publishing (2015).

Villegas, M. et al. (2016). General overview of ImageCLEF at the CLEF 2016 labs. In: CLEF 2016 Proceedings. Volume 10456 of Lecture Notes in Computer Science., Evora. Portugal, Springer.

Xu, R., Jiang, N., Lin, C., Mrachacz-Kersting, N., Dremstrup, K., & Farina, D. (2014). Enhanced low-latency detection of motor intention from EEG for closed-loop brain-computer interface applications. IEEE Transactions on Biomedical Engineering, 61(2), 288~296.

Index